The Regional Economic Impact of Technological Change

The Regional Economic Impact of Technological Change

Edited by
A. T. Thwaites and R. P. Oakey

St. Martin's Press, New York

Library of Congress Cataloging in Publication Data
Main entry under title:
The regional economic impact of technological change.
    Includes index.
    1. Technological innovations.   2. Regional economic
disparities.   3. Regional economics.   I. Thwaites, A. T.
II. Oakey, R. P. (Raymond P.)
HC79.T4R43   1985   338'.06   83-13851
ISBN 0-312-66906-2

Contents

1 Editorial Introduction

A. T. THWAITES and R. P. OAKEY
Centre for Urban and Regional Development Studies
University of Newcastle upon Tyne

1.1 Technological change and industrial growth

Why should we be interested in the regional dimensions to technological change? Technological changes have historically had a profound effect upon our civilization (Redlich, 1954-5). Even before the industrial revolution, while such changes may have been relatively rare, innovations such as the printing press, gunpowder and glass windows altered our society and, in turn, created new industries. With the onset of the industrial revolution the pace of change quickened, increasing the output per unit of labour input, and providing a wider choice of goods upon which increasing income could be spent. Once it was clear that technological change could have a positive effect on an economy, the myth postulated by classical economists that there were limits to economic growth was refuted (Mill, 1862). Although scholars have been interested in, and commented upon, the impact of innovation or technology on the socio-economic system since the time of Bacon (1625), it was not until the mid-twentieth century that an attempt was made to measure the effect of technological change upon economic growth. Solow (1957), in his pioneering work on this subject, estimated that over 80 per cent of growth observed in the United States economy between 1909 and 1949 was attributable to technical progress rather than to increasing inputs of capital and labour. Whilst his approach was subsequently criticized, later researchers such as Denison (1967) continued to attribute a substantial proportion of observed growth in the United States, West German and British economies to advances in knowledge. These reasons alone perhaps justify our interest in subnational levels and rates of technological change.

If the link between technological progress and economic advance in general is accepted, it is also evident that long-term differences in the rates of national technological progress exist. For example, over the past century, relatively few nations have been able to overcome their original economic backwardness to join the leading nations by means of rapid technological change based on industrial innovation. This is naturally a cause for concern amongst those nations outside the favoured few, while among the favoured nations there is a desire to retain, and even build upon, these original advantages in order to maintain such disparities. As Boretsky (1975) claims, the United States' economic and military strength is based upon its technological capability; a capability which, according to Boretsky, should be further enhanced. He argues that one way to ensure that the United States retains its economic position in the world is to prevent the current leakage of technology from within the country to its major competitors abroad; a process made easier through licensing. At the same time, the economies in receipt of technologies from developed Western economies are keen to exploit these advantages and, by so doing, reduce the technological and economic gap between themselves and the donor nations. In the twentieth century, and in particular in the recession that has hit the last quarter of the twentieth century, technological progress has become an extremely competitive issue between nations and groups of nations.

Although there is obvious kudos attached to leadership in technological change, the explicit pursuit of this goal is generally driven by an awareness of the broader implicit costs and benefits of lagging or leading in the 'race' to obtain technological progress. Rapid technological advance in any economy can bring benefits of increased national income and output, an increased ability to trade internationally, and the influence that accompanies technological leadership in economic terms. Technical change can improve the competitive position of firms within the economy *vis-à-vis* other foreign firms, providing an opportunity to penetrate home and international markets and perhaps displace old industries or forms of production. On the other hand, technological change is a threat to lagged nations and industries. Technological change often destroys traditional advantages which may lead to new locations of production. It can bring shifts in industrial importance where old industries are invaded by new products and old services are replaced by more modern techniques. These shifts are often reflected in terms of changed output, income, employment, skill or occupational structures, which affect overall economic performance, individual incomes, and job prospects.

It is apparent that, in a continually changing environment, it becomes necessary for nations to execute technological change, at least as fast as competitor countries if long-term economic decline is to be avoided. This is currently of great importance if, as many commentators suggest, the world is poised to experience an upsurge in economic growth based upon new technologies, emanating from the microelectronic revolution. Freeman (1985) suggests that currently distressed economies will not be revived along traditional lines, but will require new organizational and institutional arrangements if they are to benefit from the new micro-electronics era. Technology is also permitting restructuring of international corporations which will be of benefit to some world regions rather than others. As a result many governments have intervened to retain or enhance existing locational advantages for high-technology production and to off-set external technological invasion through international collaboration with sectoral and technology specific aid schemes. For example, the European Community has brought together the major European pro-ducers of microelectronic products to take part in a research programme designed to reduce the technological lead currently held by the United States and Japan. In the United Kingdom there have also been aid schemes designed to rejuvenate the specification of products (and their competitive-ness) through the encouragement of R & D work. Aid has also been available in Britain more generally to encourage the incorporation of microprocessors in both products and processes. Even in countries where there is less formal intervention in industry, appeals have been made to restrict the outflow of technology to competitors and potential com-petitors where such technologies are seen as the basis for competitive advantage.

1.2 Technological change and regional developments

These legitimate hopes and fears, based upon technological progress at the national level, are equally valid at the subnational or regional levels. Rapid technological advance within industrial firms in an area can give competitive advantages in local, national and international markets which can, in turn, lead to improved prospects of output, income, employment and occupational diversity. Similarly, innovative activities in the service sectors of the local economy can support such manufacturing innovations through business or public service provision, and thus help the local area to participate effectively in new technological developments. Conversely, regions lagging in technological developments are, without protection, likely to see their economic base eroded by external competitive forces, leading to lower output, incomes and employment. Unlike the national situation, until comparatively recently, few nations have implemented region-specific technology policies designed to overcome any perceived shortcomings. One of the fears of the executives of firms located in regions already economically disadvantaged is that the new technologies will not benefit but, rather, further disadvantage them and that the per-sistent nature of their problems of technological backwardness will either continue unchanged, or be exacerbated by further technological change in the world economy.

It is against this background that an increasing amount of research has been undertaken to explore various aspects of the regional dimension to technological change where the conclusions frequently have implications for policy at different levels of national and local government.

1.3 Subsequent chapters and their broader contexts

The chapters in this book reflect the recent work of a number of researchers concerned with the regional dimension to technological change within advanced industrial economies. Various aspects of technology are examined, ranging from research and development effort to the diffusion of new technologies in the public and service sectors of the economy. We believe that a compilation of such work is timely since the study of the impact of technological change on the prosperity of regional economies, from both theoretical and policy viewpoints, has now become established as a coherent subdiscipline within the broader subject area of regional development studies.

In Chapter 2, Morgan Thomas expresses dissatisfaction with existing conceptual frameworks and indicates ways in which we might construct a more adequate framework to achieve a better understanding of the technological change process, and its effects upon local economic development. Thomas is eminently qualified to make such judgements and recommendations because of his long-standing participation in, and awareness of, developments in this subject area. He is critical of the aggregative approach at either the industrial or geographic scales which, of necessity, tends to use generalized factors and aggregated independent variables such as labour or capital in 'explaining' regional economic development. These abstractions from reality provide little insight into the process of technological change and its relationship to regional development. He argues that considerable improvement to macroeconomic theory could be made by closer examination of their micro foundations. Only in this way will causal mechanisms rather than associations be determined. In arguing for the abandonment of aggregate macroeconomic studies of economic growth in favour of micro studies, he also suggests that there is merit in studying subsectors of the regional economy and their technologies, rather than the regional economy or its technologies in total. Such an approach avoids the aggregative method which amalgamates factors that are not conducive to individual measurement since subsequent disaggregation is fraught with difficulties.

As a vehicle for such theoretical development and greater understanding Thomas suggests a concentration on the contribution that small high-technology firms make to regional development. Once explanatory frameworks are satisfactorily developed at this limited scale, they can then be applied at a larger scale to provide a full understanding at the regional level of their complex causal relationships. Thomas suggests that if we are to obtain such an understanding, there is a need for an interdisciplinary approach to theoretical development. He goes on to spell out the complexity of the relationship between technology, enterprise characteristics, and decision-making and the local regional environment within a dynamic framework. In doing so he questions current approaches to research in this field, suggests ways in which complementary methodologies and research in other fields might be put fruitfully to use, and how many of the research problems he raises might be overcome. In this he poses many challenges to current and future researchers who seek to explain the relationship between technology and regional economic development in dynamic situations.

For research and policy purposes, and following Thomas's advice for disaggregation, it is necessary to define more clearly the term 'technological change'. Close examination suggests a number of dimensions to the term, some of which overlap one another. Schmookler (1972) suggested that the technological capacity of any economy can be defined as the accumulated body of technical knowledge weighted by the number of persons who have access to this knowledge. Technological change can therefore come about through increases in the body of knowledge *or* changes in the number of people who have access to it, or a combination of both. Furthermore, technology is not a single item, but covers a spectrum of activity, including the advancement of scientific knowledge *per se,* and the incorporation of this knowledge into products, processes and management innovations.

One extreme of the technology spectrum is generally associated with basic research in which the quest for knowledge is pursued without any specifically defined economic goal. This type of research is frequently carried out in institutions of higher education and other publicly sponsored research establishments and in the longer term may be relevant to advances of a more practical and commercial nature. The commercial enterprise, a pragmatic organization, generally concentrates upon applied research resulting in inventions which might be exploited in the marketplace or used for practical purposes within society generally. Such organizations and firms also expend considerable sums on development to translate basic ideas and inventions into marketable propositions which might include prototype and pilot developments. Innovation is generally defined as the first commercial or genuine application of some new development outside of experimentation, and it is at this stage that technological change begins to have considerable impact on the structure of economic activity changing the demands for capital, materials, labour and skills and introducing the possibility of substitution as 'consumers' make choices between competing goods within the limits of scarce resources. The products which emanate from commercial R & D are also predominantly enjoyed in the region in which the research establishment is located (Vernon, 1966; Norris and Vaisey, 1973; Oakey *et al.*, 1980).

Research has suggested that the incidence of R & D within the industrial firm may be a key determinant of the rate at which technological advance occurs (Freeman, 1982; Norris and Vaisey, 1973). While not all R & D results in 'world-beating' products or processes, there is a general correlation between R & D effort (measured in financial or employment terms) and innovative success. Thus, regions with a high incidence of R & D activity will benefit from the employment that results from successful application of R & D in the production of goods and services within enterprises.

Beyond the confines of the individual firm, an accumulation of R & D in many firms of a region produces a 'critical mass' of skilled white-collar manpower which acts as a retentive force on existing firms with a research orientation, and attracts similar firms from outside the region or abroad. Clearly, the clustering of R & D-intensive firms in specific regions (e.g. the South East of Britain or the Silicon Valley area of California in the USA) implies that, while cumulative *advantage* accrues to such areas, cumulative *disadvantage* may occur in areas from which such research-orientated growth is attracted or diverted. In addition to the agglomerative tendencies of R & D-intensive firms in particular regions, it is also common for concentrations of this type to be particularly productive 'seed beds' for the 'spin-off' of new small firms (e.g. Silicon Valley—Oakey, 1984). The addition of the new technology-based firms to local regional economies adds to diversity and strengthens the production capabilities of the local area. The importance of R & D to enterprise and local economies in terms of enhancing technological opportunities, employment and higher-order occupations has frequently been demonstrated. However, our understanding of the causal mechanisms which account for the specific locations in which we find R & D within any economy are less well developed.

In the first of the three chapters on R & D, Buswell, Easterbrook and

Morphet (Chapter 3) provide an update on the location trends of R & D within Britain, which confirms the broad pattern detected in earlier work reported by Buswell and Lewis (1970) in what is now regarded as a seminal paper on the subject. These latest results continue to show the concentration of R & D activity within the South East and core regions of England. The authors suggest that such descriptions are inadequate explanations of the regional dimension to R & D and need to be set within a broader framework of industrial geography including industrial segmentation, economic cycles and variations in the types and structures of firms involved which takes account of their position in the production system as a whole. Parts of this system in the United Kingdom are heavily dependent on the defence industry for support which therefore influences locational decision-making. The authors also remind us that in any study of the geographical location of technological activities we should never ignore the 'consumer' and the location of markets.

In Chapter 4, John Britton presents a comprehensive picture of the problems that inhibit the development of a viable, *indigenous* R & D sector in Canada nationally and subnationally, and the resultant implications for innovation and other high-technology developments. Inevitably, the issue of foreign control is central to the main arguments of the chapter. Britton indicates the pervasively detrimental effect that foreign ownership exacts on R & D indigenous to Canada. The overwhelming influence of the United States to the south tends to produce Canadian branch plants of US parents with little Canadian input to the R & D process. Such American firms generally prefer to locate R & D near to their headquarters location in the United States. This practice inhibits the development of a Canadian R & D sector, perhaps precluding the birth and growth of indigenous Canadian firms which might have a higher industrial R & D commitment. Britton considers Canadian government policies for the promotion of R & D, including tax relief and direct government funding, and the potential these have for encouraging R & D growth.

In the last of three chapters on the theme of R & D activity (Chapter 5), Malecki examines the location of public sector R & D in the United States, and the roles of these establishments in acting as nodes of high-technology R & D capacity from which new firms might spin off, and to which existing firms might be attracted. Malecki finds government-funded R & D corporately, sectorally and locationally concentrated. The local effects of such funds being dependent upon the coexistence of other factors in the area and upon the nature of input–output linkages which in the case of public expenditures, such as those related to defence, are frequently complex. Throughout the chapter, and particularly in the conclusions, Malecki suggests that federal programmes of research have little if any explicit regional policy dimension but result in a variety of regional effects.

Beyond academic institutions, it is relatively rare in industrial firms for research to be undertaken simply in the pursuit of knowledge *per se*. A more pragmatic objective, particularly within industrial and commercial settings, is the production of some new product or process. However, innovation is not limited to direct production, but also takes place in the management and organization of economic and social activities and within both the public and private sectors of the economy. It should, however, be noted that not all advances are of equal weight in terms of their economic or social value. *Fundamental* changes, such as developments in microelectronics, have a major direct or indirect influence on all aspects of our working and social lives as they diffuse through the economic system. Such innovations are subject to continuous improvements over a long period of time and form the basis for further *major* innovations designed to meet the expressed or implicit needs of society.

The origin of innovations which make a major contribution to economic activity is frequently the subject of the 'in-depth' case study, but the study of the technological capacity or advance of any defined region requires a broader approach to determine the *aggregate* impacts of *all* the innovation of a region in a given time period. Relatively few of these studies have been undertaken. At the regional level in the United Kingdom, evidence from the Science Policy Research Unit, University of Sussex, suggests that the traditionally depressed areas of the country in the North and West generate fewer major innovations than do the core areas of the country and that the proportion of substantial innovations in the periphery is decreasing over time (Townsend *et al.*, 1981). This lack of innovative activity in less favoured areas in the United Kingdom is supported by other work undertaken in the Centre for Urban and Regional Development Studies, University of Newcastle upon Tyne (Thwaites *et al.*, 1981).

If the generation of new technologies is spatially differentiated this can be compounded or alleviated through the adoption of technologies available in the world economy. The widest *spatial* economic impact of technological advance comes from the diffusion of process technologies— the wider the diffusion, the greater the impact on regional economies. It is the diffusion of process innovations that facilitates many incremental improvements in efficiency of regional industries, improvements which frequently make the products of such industries more competitive in terms of both quality and price. Those economies able to absorb new technologies to their advantage may overcome some of the disadvantages associated with the inability to generate new technologies. The risks, costs and time of participation in technological advance may be reduced while the rewards may be considerable. The rate and level of the diffusion of technology in a regional context may therefore be of considerable importance to the rate and level of economic advance.

The first chapter on diffusion by Gibbs and Edwards (Chapter 6) is based on evidence from the UK element of an international collaborative study examining the spatial and temporal diffusion of selected innovations in the mechanical engineering sectors of industry. The techniques which provided the focus for the research included the computerized numerical control of metalworking machine tools, more general applications of computers in manufacturing industry and the incorporation of microprocessors into production processes and as a subcomponent into products. Their enquiries suggest that while regional and temporal variations by region exist, other structural and influential factors in the adoptive process are at work. These include the scale of operation, corporate status and involvement in research activities at the local level. An interview survey which formed part of this research methodology permits the authors to comment on the broader characteristics of adopters and relevant non-adopters and on those qualitative factors which influence their adoptive behaviour. Finally, the influence of various aspects of public sector aid on the adoption of technology by establishments is explored.

The chapter by Rees, Briggs and Hicks (Chapter 7) also investigates the diffusion of process innovations in the American machinery industry. Using a similar methodology to that designed by Gibbs and Edwards, the research produces a number of interesting findings which tend to contradict some of the current ideas on the demise of the 'frost belt' and the rise of the 'sun belt' and their respective constituent states. It appears that in the traditional machinery industries, common in the industrial heartland of the Mid-West, it is generally the more traditional areas that are most adoptive and the 'sun belt' states that are the laggards. Rees, Briggs and Hicks conclude that it would be foolish to 'write off' traditional industries in favour of the newer high-technology sectors, or underestimate the capacity for staple industries to adapt to new, more competitive conditions in the medium term.

The final chapter in the diffusion section of the book deals with a non-industrial aspect of the diffusion process and its impact on regional prosperity. Kodras and Brown consider the diffusion of food stamps through the evolving food stamp programme in the United States. The paper offers a number of universally applicable conceptual insights on both the quality of *provision* of a new innovation and the *receptivity* of the potential adopters: essentially the respective quantity and quality of supply and demand in the diffusion processes. The research considers the varying vigour with which the programme was pursued by state agencies in various counties throughout Ohio in the Mid-West of the United States. By means of regression analysis, estimates are made of the contribution to the efficacy of the total diffusion process of factors related to the 'quality of provision' and those of the recipient population.

In the final chapter, Goddard *et al.* highlight a research area of growing importance to any explanation of regional economic development: the increasing supply of, and demand for, information, which has been prompted by the development and diffusion of the new information technologies (NIT). The impact of NIT on economic development is discussed within a conceptual framework which includes both the city and the wider region of which it is part. The authors argue that NIT provides the *opportunity* to reverse the centralization tendencies within economies observed in recent years and noted in some of the previous chapters.

However, the same technologies also permit increased centralization at a variety of spatial scales (international, national, regional, urban and suburban), and in particular within organizations such as multinational companies operating on a world scale. Thus, since the restructuring of the world's economic activity is taking place at an increasing pace, those cities or areas which will benefit most from such readjustment must either offer, or be able to develop, NIT structures necessary to realise the full potential of the firms located in, or attracted to, the urban area. The ability of cities and regions to develop advantageous structures will be deeply conditioned by existing infrastructure, industrial and institutional capacities. Goddard provides evidence of spatial variations within Europe of NIT supply and demand and goes on to assess the impact of these technologies on both manufacturing and service industries and within both large and small firms. Finally, he lists proposals for positive and co-ordinated action designed to help less favoured areas overcome their current problems: problems which reduce their capacity to participate in the benefits accompanying these technological developments but, if left unresolved, increase the likelihood of these local areas being further disadvantaged as the NIT diffusion process accelerates.

In a research area with a volatile and growing body of researchers and investigations, it is impossible to draw final conclusions in any meaningful way. Thus we offer this collection of chapters as a sample of the current work undertaken by key researchers within the field of technological change in regional development. We do not believe that the elevation of technological change to a more prominent position in attempts to explain regional economic growth is a temporary aberration. We feel sure that research into the regional impacts of technological change will continue to grow in both the industrial and service sectors as the pace of technological change continues to increase. The discussions presented in the following chapters should be considered, therefore, as an interim report on a growing subject area and in no sense the 'last word' on this theme. The value of this work lies not only in the evidence on variations in technological change between regions, but in the many fundamental issues raised to challenge future research.

References

Bacon, F. (1625). *The Essays or Counsels Civill and Morall* (London).

Boretsky, M. (1975). 'Trends in US Technology: A Political Economist's view', *American Scientist*, **63**, 70–82.

Buswell, R. J. and Lewis, E. W. (1970). 'The Geographic Distribution of Industrial Research Activity', *Regional Studies*, 4, (2), 297–306.

Denison, E. F. (1967). *Why Growth Rates Differ* (Washington, DC, Brookings Institute).

Freeman, C. (1982). *The Economics of Industrial Innovation* (London, Frances Pinter), 2nd Edition.

Freeman, C. (1985). 'The Role of Technical Change in National Economic Development' in A. Amin and J. B. Goddard, *Technical Change, Industrial Restructuring and Regional Economic Development* (London, Allen and Unwin).

Mill, J. S. (1862). *Principles of Political Economy* Vol. II (London), 5th Edition.

Norris, K. and Vaisey, J. (1973). *The Economics of Research and Technology* (London, Allen and Unwin).

Oakey, R. P. (1984). *High Technology Small Firms* (London, Frances Pinter).

Oakey, R. P., Thwaites, A. T., and Nash, P. A. (1980). 'The Regional Distribution of Innovative Manufacturing Establishments in Britain', *Regional Studies*, 14, 235–53.

Redlich, F. (1954–55). *The Role of Innovation in a Quasi-static World: Francis Bacon and His Successors* (Cambridge, Mass., Harvard University Press).

Schmookler, J. (1972). 'Patents, Invention and Economic Change', in Z. Griliches and L. Hurwicz (eds) *Patents, Invention, and Economic Change* (Cambridge, Mass., Harvard University Press).

Solow, R. M. (1957). 'Technical Change and the Aggregate Production Function', *Review of Economic Statistics*, 39, 312–20.

Townsend, J., Henwood, F., Thomas, G., Pavitt, K., and Wyatt, S. (1981). 'Innovations in Britain since 1945', Occasional Paper Series No. 16, Science Policy Research Unit, Sussex University.

Thwaites, A. T., Oakey, R. P., and Nash, P. A. (1981). 'Technological Change and Regional Development in Britain', Final Report, Centre for Urban and Regional Development Studies, University of Newcastle upon Tyne.

Vernon, R. (1966). 'International Investment and International Trade in the Product Cycle', *Quarterly Journal of Economics,* **80**, 190–207.

2 Regional Economic Development and the Role of Innovation and Technological Change

MORGAN D. THOMAS*
University of Washington, Seattle

1. Introduction

Long-run processes of regional economic development are extremely complex phenomena, and currently we are unable to provide an adequate explanation of the conditions under which economic activities grow and decline and change their locations, industrial composition and relationships within a specific region. This situation is exacerbated by our inability to provide, over time, the kind and quality of information needed to evaluate the factors and conditions responsible for the spatial, economic, technical, organizational and other characteristics of the economic activities in a region.

Traditional theoretical approaches to regional development have generally 'explained' dynamic economic activity patterns at a high level of industrial aggregation and at a macrogeographic scale. Because the main theoretical thrust has been at these macro levels or scales, the resulting 'explanations' tend to have a mechanistic quality. They are also based mainly on statistical relationships between regional economic development and aggregate independent variables such as labour, technology, education, taxes and capital. Moreover, these same theories have, among other things, failed to explain the spatial tendencies of a given region's industries, firms and establishments, and the influences on these tendencies that are generated by the region itself.

In the 1950s traditional approaches were challenged and considerable impetus to theorizing about the process of economic growth and structural change took place through the use of a more disaggregated approach. The broad primary, secondary and tertiary sectoral categories of national economies were disaggregated into a greater number of individual industries. In this way it was possible to see much more clearly that industries varied in their rates of growth over time and geographic space. As a consequence of this disaggregation the dynamic explanatory elements were more

* I wish to thank Gunter Krumme and J. Scott MacCready for their helpful comments. This material is based partly upon work supported by the National Science Foundation under Grant No. SES-8411682. Any opinions, findings, and conclusions or recommendations expressed in this publication are those of the author and do not necessarily reflect the views of the National Science Foundation.

clearly revealed, although the explanatory task obviously grew. For example, Perroux's (1955) articulation of the growth pole theory ex-emplifies the utility of such a disaggregative approach. Considerable merit was also seen in the study of growth processes, not only in finely disaggregated industries or groups of enterprises, but also within the organizational structure of the firm itself at plant or establishment level.

Thus, several authors have argued that a considerable improvement in our macroeconomic theories could result from a strengthening of their microfoundations (Schelling, 1978; Weintraub, 1978; Rosenberg, 1982; Elster, 1983). Of course:

> The search for *micro-foundations* . . . is in reality a pervasive and omnipresent feature of science. It corresponds to William Blake's insistence that 'Art and Science cannot exist but in minutely organised Particulars'. To explain is to provide a mechanism, to open up the black box and show the nuts and bolts, the cogs, the wheels of the internal machinery. [Here the term 'mechanism' should be understood broadly, to cover intentional chains from a goal to an action as well as causal chains from an event to its effect] (Elster, 1983: 23-4).

In other words, a better understanding and explanation of the process of regional economic development will result from the articulation of an appropriate causal mechanism for these processes. However, our efforts to discover these causal mechanisms are to some degree hampered by the scale of the problem.

Because we are dealing with an extremely complex phenomenon there is therefore merit at this stage in using a more simplified disaggre-gative approach when theorizing about the process of regional economic development. For example, instead of dealing initially with the whole regional economy, conceptual and empirical work could be focused on a set of firms which belong to a particular set of industries, such as high-technology industries. Similarly, technology, which in an aggregate sense has received considerable attention as a very important explanatory factor in contemporary macroeconomic regional growth theories, can also be disaggregated into a number of related elements and dimensions such as inventions, product and process innovations, and the transfer and diffusion of technology (Thomas, 1975a). In turn, the role and influence of these elements and dimensions of the technology factor can be considered and evaluated within intentional chains which cover behavior from goal to action in high-technology firms. They may also be considered as elements in a causal chain linking an event such as a product innovation to its effect, such as bringing about growth and diversification in a firm's output or a change in its location.

In taking this disaggregative and selective approach the following paper presents a discussion of current explanatory notions and tentative hypotheses found in the literature concerning the causal relationships between innovation and the process of regional economic development. Hopefully this partial approach will contribute to the development of better explanatory frameworks which can be applied at the larger scale and help to provide greater understanding at all levels of these complex causal relationships. In particular, a better comprehension of the role of innovative high-technology firms in the processes of regional economic growth and structural change in itself should:

(a) improve individually and collectively the theories of innovation, the firm and regional growth and development: and
(b) provide more effective guidance for the design and use of innovation-oriented regional economic development policy instruments in the United States and in similar countries.

2. Macroeconomic growth theory, innovation and regional economic development

In the early 1950s, the dominant explanatory variable for the processes of long-run economic growth and structural change was capital. The Capital Stock Adjustment Theory and its many direct descendants provided the prime theoretical guidance for those interested in providing a coherent explanation for economic growth and change in the composition of industries. Before the end of the decade, however, the scene was set for a major 'take-over' by technical innovation as the 'new causality' (Pavitt, 1979). Major American contributors in this process were Kuznets (1952), Solow (1957), and Denison (1962). Thus, during the latter part of the 1950s and early 1960s, certain mechanical macroeconomic relationships and associations between technical innovation and economic growth and structural change were introduced explicitly into neo-classical growth theory.

However, the continuing dominance of neo-classical growth theory, albeit modified to include the concept of technical change, still fails to provide an appropriate articulation of the dynamism of the innovation time path because this type of theory does not provide an *adequate* understanding of the process of innovation within a temporal framework. Furthermore, it does not produce coherent explanations of the dynamic causal impacts of innovation on rates and direction of economic growth and on changes in the composition of industries within regional systems (Thomas, 1975a). However, attempts to understand the macroeconomic structural changes taking place over the last decade at global, national and subnational scales have revived considerable interest in the study of cyclical forms of long-term patterns of economic growth with primary attention focusing on Kondratiev cycles (40–50 years) and Kuznets cycles (15–25 years) (Nijkamp, 1982a). Normally the cyclical pattern of long-waves in a capitalist economy includes a number of stages from trough to trough: take-off, rapid growth, peak followed by decline. Attempts have been made recently to relate these long-term wave patterns of economic growth in industrialized countries to cyclical patterns in innovation and diffusion processes. In particular, concern has focused on those processes based upon some fundamental technological change such as the introduction and development of the railway system (Clark, *et al.,* 1981; Kleinknecht, 1981; Mensch, 1979; Nijkamp, 1982a). These technological cycles to a large degree mirror the economic cycle: fundamental innovation, swarming of supplementary innovations, diffusion, maturity and, inevitably, decline. The growth and structural change relationships of these long-term wave patterns, however, do not appear to have been studied to any great extent in subnational spatial contexts largely because of a lack of data (Nijkamp, 1982b). We are therefore ignorant of whether or not the full cycle or only parts of these cycles are experienced in each area and what the implications of such experiences might be for economic development in specific local areas.

In his recent commentary on economic studies of innovation and technological change, Malecki (1983, 98) observes that a number of economists, generally recognized as among the foremost researchers on this topic, such as Gold (1977, 1979, 1981), Nelson (1981), and Nelson and Winter (1974, 1978), also express dissatisfaction with the aggregate nature of research on innovation and technological change. These and other equally well-known economists such as Nabseth and Ray (1974) and Rosenberg (1982) have called for, and used, more realistic disaggregate approaches in their research, based on individual firms. However, in contemporary America, despite the shortcomings in conceptualization and research noted above, the panacea for revitalizing depressed economies at subnational and national levels appears to be the high-technology industry 'quick fix'. 'High-technology' firms and industries are often defined by characteristics such as relatively high R & D expenditures and growth rates, innovativeness, high wage structures and competitiveness in world markets. Such firms and industries, therefore, promise those who live in economically stagnant or declining areas a more attractive and better future. In recent times much thought and effort have been directed towards the development of public policies designed to attract, establish and nurture the growth of innovative firms and industries in such areas.

Indeed, there is ample evidence from many parts of the world of the articulation of a new approach to national and regional economic development policy which emphasizes "innovation promotion' and the explicit use of technology-oriented policy instruments (Ewers and Wettman, 1980; Rothwell and Zegveld, 1982; Hill and Utterback, 1979; Nelson, 1982; Botkin, Dimancescu and Stata, 1982; Sirbu *et al.*, 1976; Steed, 1982; US Congress, Joint Economic Committee, 1982). The complex nature of problems to be addressed perhaps require inputs to policy design from more than one discipline. However, while economists rarely deal explicitly with the spatial dimensions of the processes of innovation and economic growth and structural change, geographers, conversely, have rarely dealt in depth with the economic dimensions of these processes. Members of each discipline tend to cluster their research questions concerning innovation and technical change around one approach, e.g. economic or spatial. Research studies in this field, consequently, tend to be characterized by their general lack of recognition of the need to integrate the necessary contributions made by members of many disciplines (NSF, 1983).

Recently, however, a number of economists and geographers have made important interdisciplinary contributions focused on the relationships between technology and regional economic development. Steed (1982) examined small and medium-size innovative firms in Canada. Spatial diffusion of innovations involving a number of high-technology industries was surveyed by Thwaites, Oakey and Nash (1981) in Britain, and by Rees, Briggs and Oakey (1983) in the United States. Research and development cycles, investment cycles and regional growth in Britain and American industries have been studied by Oakey (1983). Regional economic growth effects and location patterns of research and development establishments in the United States have been explored by Malecki (1979, 1980a, b, c, 1981, 1982). Brugman (1983) added a spatial perspective to the study of technological innovation in technology-intensive industry. Interregional locational change in the United States semi-conductor industry was examined by Harrington (1983), while Saxenian (1981) documented some intraregional spatial consequences of industrial restructuring in the Santa Clara County (Silicon Valley), California. Evidence of spatial tendencies in high-technology industries in the United States has been provided by Glasmeier, Hall and Markusen (1983) at Berkeley, and Armington, Harris and Odle (1983) at Brookings.

However, the conclusion emerging from this review of theory, research and practice is that we still lack an adequate conceptual framework within which we might fully understand processes of technological development and how they impact upon local economic development. The remainder of this paper will be devoted to the development of such a framework.

3. The region and innovative firms: a dynamic framework

In this paper we are essentially concerned with explaining the economic development role of innovative firms in a regional or spatial context, specifically focusing for illustrative purposes on the contribution of innovative high-technology firms to economic development. The conceptual framework articulated will, therefore, need to provide guidance in the search for the dynamic interdependent relationships between the region ('the regional factor') and the innovative high-technology firms ('the innovation factor'). Causal sequences within these relationships between the region and the firms provide an important part of the mechanism which explains such events as the location of the high-technology firms in the region, as well as changes in their size and output composition. In addition, causal sequences between firm management decision-making behavior and firm-level processes of innovation, growth and diversification of output within the regional high-technology firms are expected to have a number of important spatial dimensions or implications.

Growth in output and/or changes in the composition of the outputs of individual innovative firms may, for example, be reflected in the construction of new plants and/or the purchase of existing plants within a study region and/or in other regions. These changes may well initiate or be initiated by changes in the organizational structure and intra-firm innovation as well as in the technology diffusion patterns of the individual firms. In addition, output and product composition changes in the innovative firms over time would be expected to require continuing changes in the set of regions from which inputs originate for such firms, as well as changes in the set of regions that are the destinations of their products. These dynamic spatial dimensions of growth and change in innovative firms have not been considered and used adequately in the body of theory which explains the process of regional economic growth and structural change.

3.1 *The regional factor*

The 'regional factor' is a neglected and little understood explanatory variable in regional economic development theories (Thomas, 1975a; Thomas and LeHeron, 1975; Martin *et al.*, 1979). For present purposes this meta-variable represents the net influence of a region's human and non-human attributes on the processes of economic growth and structural change in the region. These attributes include, for example, the region's: population; work-force; occupational structure, industries, firms and establishments of various kinds and sizes producing a variety of products (raw materials, goods and services); urban and transportation systems; social and economic infrastructures; sets of laws, government policies, incentives and regulations; institutional structures related to education and social and commercial activities; minerals, timber, water, climate and other natural resources or elements; and its location relative to other regions and places.

These endowments of a region can be categorized according to the various attributes they possess of relevance to the particular phenomena under investigation; for example, educational institutions can be classified according to the industries or the occupations they serve or the specific skills, scientific or technical training or experiences they impart. Banking institutions may be classified as to whether or not they provide 'venture' capital; and the region's economic climate may be categorized according to its suitability for the high-technology firms already established or considering location in the region. The region's pertinent characteristics and endowments need to be examined and explained in both intra- and interregional spatial frameworks.

Comprehensive empirical studies of the explanatory role of the regional factor in regional economic development processes are rare. Martin *et al.* (1979) examined the systematic effect of selected regional endowments on the interregional diffusion of innovations in Canada. Goddard (1983), Thwaites, Oakey and Nash (1981), and Thwaites, Edwards and Gibbs (1982) have also observed interregional variations in the innovativeness of firms in certain high-technology industries and they have suggested explanatory relationships between these variations and regional differentials in selected attributes of regions in Great Britain. These studies have used aggregate statistical associations to explain the relationships between selected elements of the regional factor and the spatial distribution patterns for innovations and innovative activity.

There are important theoretical and policy reasons for departing from this aggregate level to achieve a greater understanding of the causal mechanism underlying these relationships between the regional factor and the processes of economic development in a region in general, and the processes connected to the location, growth and structure of its innovative high-technology firms in particular. One available and important source of insights concerning the causal mechanism underlying the relationship between a region's environmental factors and its technology-based enterprises is the literature on entrepreneurs and entrepreneurial activities, the entrepreneur frequently being a key figure in technological change.

Over the last fifteen years there has been great interest shown in research on intense concentrations of entrepreneurial activity such as those found in the Silicon Valley, California; Austin, Texas, and along Route 128 around Boston, Massachusetts (Bollinger, Hope and Utterback, 1983; Cooper, 1971; Roberts, 1968, 1980; Roberts and Wainer, 1971). Research on these spatial concentrations of entrepreneurial activities, however, has been wisdom-based and lacks the guidance of a sound theoretical framework. Current knowledge of the role of the regional factor's influence on entrepreneurial activity is primarily based on case histories and ancedotal evidence and there is a great lack of empirical verification. The literature is, however, rich in insights concerning the relationships between a region's human and non-human endowments and attributes and entrepreneurial activities (Bruns and Tyebjee, 1981; Cooper, 1970, 1971; Cooper and Komives, 1972; Evan, 1976; James and Struyk, 1975; Meyer, 1978; Pennings, 1980; Vesper and Albaum, 1979).

From a study of this literature one may derive expectational propositions. For example: one would expect that the set of critical endowments and attributes of the region perceived to be very important to an individual high-technology firm will be a different set in small indigenous plants as compared to the set for a similar branch of a non-indigenous firm. For example, the availability of appropriate venture capital in the region would be expected to be far more important for the indigenous establishment than for the branch plant of a non-indigenous firm. The latter firm might well use its own investment resources generated within the firm or obtained from well established sources from outside the region.

New firms *in radically different product industries* may be expected to have a different set of critical regional environmental factors (endowments and attributes) that have a major influence on their initial location decision. This will tend to be the case also with respect to subsequent location decisions such as deciding whether or not to move elsewhere stay at the same location or to close down. This expectation is based on the fact that firms in different industries produce products that require different kinds and/or combinations of inputs from different locations and generally serve different markets. The region's role as a supplier of various kinds of tangible and intangible inputs and its role as a market for the outputs from firms in different industries would be expected to vary more in establishments of similar size across industries than would be the case for similar establishments in the same industry. One expects also that, for the same establishment, its perception of the influence or significance of specific regional environmental factors with respect to location decisions or decisions related to its growth and product, will tend to change over the life of its product(s). The establishment's perception of specific regional environmental factors will also tend to change in response to organizational changes in the establishment and/or parent firm (Thomas, 1975b).

The explanatory role of the regional factor with respect to spatial, scalar and structural changes in innovative high-technology firms in a region needs to be considered in a temporal framework because the nature and influence of the region's endowments and attributes change over time and geographic space. Over time, changes in, for example, the spatial, economic, organizational, technological and market characteristics of the region's high-technology firms will also have a dynamic impact on the nature and influence of the regional factor. This appears to be an interdependent, interactive dynamic relationship between the region and its specific set of high-technology and non-high-technology firms.

3.2 *Innovative high-technology firms*

A useful organizing framework for a discussion of technical change in innovative firms is provided by Schumpeter's notion of 'behavioural competition':

> It is not . . . price . . . competition which counts but the competition from the new commodity, the new technology, . . . the new type of organisation . . . competition which commands a decisive cost or quality advantage and which strikes not at the margin of the profits and the outputs of the existing firms but at their foundations and their very lives. (1950, 84–5.)

Thus innovations in the form of a new commodity or product, new kind of process, new form of organization, and new management practice, *provide innovators or entrepreneurs* and innovative firms with attributes which enhance their effectiveness in various forms of 'behavioural competition' or 'creative destruction' behavior. Innovations provide the firm with a 'competitive edge' or 'comparative advantage' over firms in the same industry and same market. One may also infer that the viability and growth of a firm will decisively depend on its innovative capabilities. Furthermore, to the extent that innovative behaviour and capability is unequally distributed among firms in an industry, one may hypothesize that innovation contributes to different growth rates among firms (Kohn and Scott, 1982; Kamien and Schwartz, 1982).

In the Schumpeterian framework the innovative capabilities of a firm are dependent on the entrepreneurial functions its entrepreneur(s) performs as distinct from the managerial function (Schumpeter, 1971, 35). The nature of this entrepreneurial function 'only shows up within the process of innovation' (ibid.). The entrepreneur, at different times, may carry out managerial functions, but not all individuals who carry out managerial functions are able to carry out entrepreneurial functions. Schumpeter differentiates, as did Usher (1971), between those *unlearned acts of insight* which characterize entrepreneurial functions and those *learned acts of skill* which characterize managerial functions.

It seems, however, that this distinction is becoming somewhat blurred by the fact that both entrepreneurs and managers in modern high-technology firms are increasingly dependent on institutionalized research and development (R & D) activities and investments to sustain invention and innovation. This is understandable in view of the growing scientific character and complexity of technology, and occurs despite the fact that important inventions and innovations are still made by private inventors, individual production engineers and production workers. Nevertheless, during the last seventy years most large firms in industrialized countries have established their own full-time specialized R & D sections and departments. In high-technology industries R & D inputs are much more important to the continued successful operation of their constituent firms than is the case for firms in other manufacturing industries (US Congress, Joint Economic Committee, 1982, 4). R & D activities and expenditures seem to be directly, positively and strongly related to innovation, output growth and productivity increase at industry and firm levels (Brinner and Alexander, 1977; Freeman, 1982; Nadiri and Bitros, 1979; Mansfield, 1981; Link and Long, 1981).

However, notwithstanding the recognition of R & D as important to the development of high-technology industries and of innovation within industry more generally, the time path for a specific new product *remains* poorly articulated in the literature on growth theories (Thomas, 1983). There is merit, however, in conceptualizing an innovation time path or cycle as extending through two stages. The first stage begins when the attributes of the embryonic new product are recognized and defined as an invention, and it ends with the act of innovation—the first commercial production of the new product. The second stage concerns the period of commercial production when the new product may be modified by subsequent product innovations and when additional process innovations may affect the economic efficiency of the new product's production technology. However, not all firms in the new product industry will necessarily carry out innovative activities throughout the innovation cycle of the new product. In addition, the innovator firm for the new product need not be, and usually is not, the first innovator for all subsequent product and/or process innovations.

Of considerable interest, therefore, in this conceptualization is the identification by Usher (1971, 48) of distinctive steps in the process of invention and innovation. It is useful to view these steps as groups of related non-routine decisions which occur along an innovation time path. They are dimensions which involve necessary *unlearned acts of insight* and *learned acts of skill* respectively by the firm's entrepreneurs and managers. These non-routine decisions may affect profoundly not only the life cycle of the innovation, but the future well-being of the innovative firm. For small or new firms such decisions often have a virtual 'life or death' importance.

Over the innovation life cycle the innovator firm encounters distinctive decision steps which involve novel situations and conditions which reflect high levels of uncertainty. Clear-cut procedures for the firm indicating how to deal appropriately with such situations and conditions would, therefore, not be expected. Decisions such as whether or not the firm should 'set the stage' and move towards the primary invention phase or move to the critical revision and development or innovation phase are major decisions and normally one would not expect them to be made in a routine fashion using only routine procedures. It is, therefore, important to understand the behavior of the firm when it needs to make non-routine decisions.

The behavior of firms in an industry faced with non-routine decisions regarding their future is influenced by the distilled memory of the history of each member firm. Some of the experiences from the past and present are represented, appropriately or not, in the current decision rules of the firms. When making decisions regarding their future well-being, firms attempt to synthesize past and present flows of information, generated within the firm and from external sources, concerning the present and future decision environments of the firms (Nelson and Winter, 1977; Thomas, 1980; Piatier, 1981).

The decision environment or selection environment of a firm is determined partly by conditions within the firm which influence its behavior.

The characteristics and behavior of other firms in the firm's industry also partly determine the nature of a firm's decision environment. In addition, conditions external to both the firm and its industry play an important role. Such external conditions include product demand and factor supply conditions and those that influence other information flows to the firm. Inevitably there are inherent weaknesses in the information flows from the past and present and future decision environments of the firm as well as deficiencies in the firm's ability to process, evaluate and utilize the information correctly. These are only a few of the factors which, for the firm, introduce varying degrees of uncertainty with respect to the outcome of a non-routine decision regarding its future.

An innovative firm builds into the organization a set of ways of doing things and ways of determining what to do. These organizational routine behaviors, however, will not be sufficiently appropriate for dealing with unprecedented situations or new and changing conditions in the decision environment of the firm. In such situations the firm's well-being will be dependent on how effectively it can adapt or replace current organizational routines. Consequently, it would be expected that the firm would initiate organizational search and evaluative activities which might lead to the modification, or to drastic change, or the replacement of current routines. Of course, many of these search and evaluative activities are, in effect, organizational routines which manifest a stochastic character (Nelson and Winter, 1982b).

These three notions—decision environments, organizational routine, and search—provide a foundation for the development of a conceptual framework for understanding the behavior of a firm when it deals with non-routine decisions such as those related to innovative activities. For example, one may explore the use of this conceptual framework in seeking a better understanding of the non-routine location decisions that are made by firms as a consequence of their innovative activities. In addition the framework may have utility in seeking a better understanding of the behavior of the firm as it deals with the economic organizational dimensions of the innovation process. The technical success of the new product cannot ensure the commercial success of the innovation if the right kind of organizational structure and management practices are not in place (Slater, 1980). The framework may also be used in the study of non-routine decisions related to the development of R & D strategies and projects during the first and second stages of the innovation life cycle (Stead, 1976; Horesh and Kamin, 1983).

In many manufacturing industries (OECD, 1971; DeKluyver, 1977) and in agriculture (Binswanger and Ruttan, 1976), there appears to be support for the hypothesis that there is a precommitment on the part of a firm to advancing a technology in a certain direction referred to as its natural trajectory. A natural trajectory is associated with a particular technology (embodied or disembodied), or technical regime. For example, in the 1950s the Boeing 707 jet defined a particular technological regime. Subsequently, for three decades in different aircraft firms, each with its own natural trajectory, engineers have introduced innovations designed to exploit the potential for the technological regime of the commercial jet aircraft first conceived by Boeing.

The concepts of natural trajectory and technical regime seem to be related respectively to the concepts of firm-specific technology and firm-system-specific technology (Hall and Johnson, 1970; Thomas, 1975b). A firm's natural trajectory or firm-system-specific technology is an important source of monopoly (usually transitory) profits for the innovative firm. It is hypothesized that the firm will, therefore, tend to be hesitant to change radically its natural trajectory and its R & D strategies related to this trajectory. The flexibility of the firm-system-specific technology will tend to influence the form and breadth of a particular firm's natural trajectory or its range of products.

Underlying this conceptualization is the major thesis that innovations may provide a competitive edge and increased growth prospects for innovative firms and industries. In addition, the presence of innovative industries, or their component firms or establishments, in a subnational region, enhances both the competitive position and economic growth prospects of the region. Nevertheless, in light of what has been said earlier concerning the uncertainties and complexities intrinsic to the innovation process, innovative effort is not a certain means of achieving economic success in firms and regions. The majority of innovations are commercial failures (Rothwell, 1977, 41). An innovation may be a technical success and a commercial failure in the innovator firm but a commercial success in the 'imitator' or 'counter-puncher' firm (Hoffman, 1976).

In the development of this framework attention is concentrated on innovation activity within a firm and on the growth and regenerative implications of this innovation activity for the firm. However, not all innovations that are incorporated in a new product over its innovation time path are the product of the innovator firm (Von Hippel, 1976). Other innovative firms frequently, if not usually, provide process innovations and intermediate product innovations for the innovator firm during the first and especially the second stage of the new product innovation life cycle referred to above. There is, therefore, an important interdependence between innovative firms in different industries and at different locations.

In addition, the inter-firm and inter-industry flows of innovations embodied in new products and processes inject into the economy what are thought to be very important input and output quality multiplier effects (Thomas, 1969). They are also believed to have significant impacts on the growth of productivity in the firms and industries connected to this intra- and interregional innovation input–output flow network (Nadiri and Bitros, 1978; Scherer, 1982). Innovations contribute to the competitive edge and long period growth of firms through: (1) the creation of new products (and processes) and (2) through economic efficiencies in production. The literature on industry and product life cycles provides useful insights in dealing with this question (Abernathy and Utterback, 1978; Gort and Klepper, 1982; Vernon, 1979).

Product and industry life cycle notions are speculative, however, and are based on information obtained from studies of a limited number of high-technology industries. Constructive criticism regarding life-cycle frameworks (Walker, 1979; Rothwell, 1976) have improved their specification and application. Research on these and related concepts continues and significant refinements are being reported in the literature. The Product Life Cycle, among other things, provides an important framework for examining changing labour–capital relationships and location strategies over the life of related streams of products (Thomas, 1975; Markusen, 1983). The spatial dimensions of these dynamic labour–capital relationships have been conceptualized primarily in macrogeographic terms. For example, over the life of a product, the spatial tendencies of its production facilities are perceived to result in a movement of these plants towards domestic or foreign regions where labour costs are lower. These spatial tendencies are viewed as one manifestation of a growing substitution of capital for labour and an increasing deskilling of the labour component associated with the production process as the product moves from the new product innovation stage to the mature standardized stage.

To date, there appears to be little conceptualization and no empirical study of the spatial dimensions of the product life cycle at the microeconomic level of the firm. Further conceptualization and empirical study is necessary for a clearer understanding of the causal explanatory relationships between innovative high-technology firms and their spatial tendencies, and patterns of growth and product diversification. There is, for example, a need for the development of a conceptual framework which will provide guidance for studying the dynamic spatial relationships of innovative firms over: (1) the pre-commercial stages of a new product life cycle, and (2) the various subsequent stages of the life cycle for a commercial new product or family of new products (Thomas, 1980, 1981a, b, 1983).

For firms that carry out research and development activities which result in new product innovations there are a variety of spatial relationships associated with these activities such as locational choices for: (a) R & D activities, (b) bench and pilot-plant testing of a potential new product, (c) initial commercial production, and (d) possible additional branch plants. Clearly the spatial patterns associated with the pre-commercial stages and the commercial stages of the life cycle of a new product will be different in single as compared to multiplant firms. Similarly, spatial variations are to be expected in the input–output linkages and organizational structures and linkages of individual firms over the different stages of the product life cycle. The changes that occur in the technological, economic and organizational conditions in environments internal and external to the firm, over the product life cycle, will tend to induce changing spatial as well as growth and product diversification responses by the firm.

Product life cycles are based on the assumption that the commercial life of a specific new product (single product industry) is finite. Product enhancement and product changes may well extend the commercial life. Nevertheless, sooner or later the product's life ends. The connection between this life cycle phenomenon and the concept of secularly declining industries (single and multiple product) is evident. Regions and firms are faced with the regenerative problem if their economic viability is to continue. They must be able to introduce new products which hopefully will be commercial successes. We do not know the relevant time path, however, during which the regenerative problem is real, for a particular region or firm, but we do know that the firm's life cycle need not coincide with the life cycle of one of its products. The expansion path of a commercially successful product innovation tends to follow the general S form. Factors such as the degree of 'newness' of the product, competition from substitutes, and changes in consumption functions of purchasers influence the precise form of the specific product expansion path (DeKluyver, 1977; Goldman, 1982).

Utterback and Abernathy (1975), in their dynamic model of innovation, specify that initially after the commercial entry of the new product (Stage II of the innovation life cycle), innovative firms, in the new product industry, concentrate on innovative activities which hopefully will enhance the 'new' product in terms of its quality, design, strength, etc. In time, within the new product industry, innovations which increase the economic efficiency of the production of the 'new' product will tend to dominate those innovations that contribute to product enhancement. If the product remains in production long enough, at some point product and process innovation ceases or virtually ceases (Utterback and Abernathy, 1975). Subsequently, factors other than the technology factor tend to have a greater influence over the commercial viability of the product. The concept of a 'product–process matrix', articulated in this model, has been used in the development and assessment of corporate strategies (Hayes and Wheelwright, 1979a, b; DeBresson and Townsend, 1981).

The product life cycle notion suggests that initially the firm's competitive position in the new product market is primarily influenced by the qualitative characteristics of its product. Demand for the new product tends to be relatively price-inelastic and unit costs relatively high. Over time, embodied process innovations contribute to a significant improvement in the cost efficiency of the new product production and to the productivity of the innovative firm. These contributions subsequently tend to result in reductions in the price and/or improvements in the quality of the product.

4. Concluding statement

At the conclusion of this chapter it is evident that there is great diversity and intensity in the current research effort directed towards the study of the role of innovation and technical change in economic development. A review of theory, research and practice suggests that considerable progress has also been made during the last decade and a half. Nevertheless, it is clear that many weaknesses remain in this work. It was shown in the discussion of this topic that a particularly persistent and important weakness of contemporary theory is that it provides a very inadequate understanding of the mechanism which explains how and why innovations influence the process of long-run economic development. This body of theory provides an even more inadequate understanding of this process in a spatial or regional context. The conceptual framework presented in the main section of this paper was developed to address briefly and selectively this particular weakness in regional economic development theory.

This conceptualization provides a way of linking a number of concepts which are associated with the processes of innovation, technical change and regional economic development. Key dynamic relationships were identified between the region (or the meta-variable 'regional factor') and innovative high-technology firms (the 'innovation factor.). The evaluation and assessment of these relationships revealed that there is an urgent need to know more about both the nature and economic development significances of the dynamic attributes of a region, and their relationship, over time, with a changing set of entrepreneurs, innovations and innovative high-technology firms.

The Schumpeterian concept of 'behavioral competition' was shown to be a most effective organizing framework for both discussing and evaluating the analytical utility of such behavioral concepts as entrepreneurial innovation and innovative high-technology firms, and understanding the process of 'creative destruction'. Furthermore, it helped to direct attention towards the 'unlearned acts of insight' performed by innovators and 'learned acts of skill' carried out by both innovators and managers. Innovation, investment and location decisions were viewed as significant 'non-routine' decisions and the processes involved in making them clearly need further study and greater understanding. The theory of heuristic search provided useful guidance for these purposes because it was shown that through the use of heuristic strategies we may come closer to identifying the 'non-routine' elements of problems which require 'non-routine' decision-making and 'non-routinized' solutions.

The conceptual framework presented in this paper underscores the various uncertainty conditions facing the firm in the pre-commercial and commercial stages of the innovation life cycle of a new product. It shows how vitally important it is that we attain greater understanding of how and why innovative firms behave as they do when encountering these various conditions of uncertainty. Information search activities in its internal and external environments, usually involving considerable costs, are clearly crucial for the innovative firm. The value of using conceptual elements, for example, the natural trajectory and technical regime, and industry and product life cycles as organizing frameworks was also demonstrated. These frameworks were shown to be helpful when seeking better explanations for the product-process relationships as well as other relationships in the processes of innovation, growth and structural change in the firm.

The conceptual probing of the microfoundations of regional economic development carried out in this paper suggests that there is merit and a need to test the validity and usefulness of the conceptualization and its constituent components in studies of innovative high-technology firms in specific regions. This overall framework does not only identify a significant and challenging research agenda; it also provides a number of suggestions as to how some of these research problems may be addressed. It is well to remember, however, that to explain fully the processes of long-run regional economic growth and structural change requires the creation and articulation of a much larger and more complex mechanism than that outlined in this chapter. Clearly, there are many worthy challenges ahead for all who wish to view and understand what is inside the 'black box' of regional economic change.

References

Abernathy, W. J. and Utterback, J. M. (1978). 'Patterns of industrial innovation,' *Technology Review,* **80** (June/July), 40-7.

Armington, C., Harris, C., and Odle, M. (1983). 'Formation and growth in high technology businesses: a regional assessment' (Washington DC, The Brookings Institutions), September 1983, (mimeo).

Binswanger, H. D. and Ruttan, V. N. (1976). *The Theory of Induced Innovation and Agricultural Development* (Baltimore, Johns Hopkins Press).

Bollinger, L., Hope, K., and Utterback, J. M. (1983). 'A review of literature and hypothesis on new technology-based firms,' *Research Policy,* **12**, 1-14.

Botkin, J., Dimancescu, D., and Stata, R. (1982). *Global Stakes* (New York, Harper and Row).

Brinner, R. and Alexander, M. (1977). *The Role of High Technology Industries in Economic Growth* (Cambridge, Mass., Data Resources, Inc.).

Brugman, B. L. (1983). *A Spatial Perspective on the Process of Technological Innovation in Technology-Intensive Industry,* unpublished Ph.D. dissertation, University of Washington.

Bruns, A. V. and Tyebjee, T. T. (1981). 'The environment for entrepreneurship' in K. H. Vesper *et al.* (eds), *The Encyclopedia of Entrepreneurship,* Ch. XVI, pp. 188-315.

Clark, J., Freeman, C., and Soete, L. (1981). 'Long waves, inventions and innovations,' *Futures,* **13**, 308-22.

Cooper, A. C. (1970). 'The Entrepreneurial Environment,' *Industrial Research,* September.

Cooper, A. C. (1971). 'Spin-offs and technical entrepreneurship,' *IEEE Transactions on Engineering Management,* **18**, 2-6.

Cooper, A. C. and Komives, J. L. (eds) (1972). *Technical Entrepreneurship: A Symposium* (Milwaukee, Center for Venture Management).

DeBresson, C. and Townsend, J. (1981). 'Multivariate models for innovation—looking at the Abernathy–Utterback model with other data,' *Omega,* **9**, 429-36.

DeKluyver, C. A. (1977). 'Innovation and industrial product life cycles,' *California Management Review,* **20**, 21-3.

Denison, E. F. (1962). *The Sources of Economic Growth in the United States and the Alternatives Before Us* (New York, Committee for Economic Development).

Elster, J. (1983). *Explaining Technical Change* (Cambridge, Cambridge University Press).

Evan, W. M. (1976). *Organizational Theory: Structures, Systems, and Environments* (New York, John Wiley).

Ewers, H. J. and Wettman, R. W. (1980). 'Innovation-oriented regional policy,' *Regional Studies,* **14**, 161-79.

Freeman, C. (1982). *The Economics of Industrial Innovation* (Cambridge Mass., MIT Press).

Glasmeier, A. K., Hall, P. G., and Markusen, A. R. (1983). 'Recent evidence on High-Technology Industries' Spatial Tendencies: A Preliminary Investigation,' Working Paper No. 417 (University of California, Berkeley, Institute of Urban and Regional Development).

Goddard, J. B. (1980). 'Technology forecasting in a spatial context,' *Futures,* April, 90–105.

Goddard, J. B., *et al.* (1983). *Technological Innovation in a Regional Context: Empirical Evidence and Policy Options,* Discussion Paper No. 55 (University of Newcastle upon Tyne, Centre for Urban and Regional Development Studies).

Gold, B. (1964). 'Industry growth patterns: theory and empirical results,' *Journal of Industrial Economics,* **13**, 53–73.

Gold, B. (1977). 'Research, technological change and economic analysis: a critical evaluation of prevailing approaches,' *Quarterly Review of Economics and Business,* **17**, 7–29.

Gold, B. (1979). *Productivity, Technology and Capital* (Lexington, Mass., D. C. Heath).

Gold, B. (1981). 'Technological diffusion in industry: research needs and shortcomings,' *Journal of Industrial Economics,* **29**, 247–69.

Goldman, A. (1982). 'Short product life cycles: implications for the marketing activities of small high-technology companies.' *R & D Management,* **12**, 81–9.

Gort, M. and Klepper, S. (1982). 'Time paths in the diffusion of product innovations,' *The Economic Journal,* **92**, 630–53.

Hall, G. P. and Johnson, R. E. (1970). 'Transfers of United States aerospace technology to Japan' in R. Vernon (ed.), *The Technological Factor in International Trade* (New York, Columbia University Press).

Harrington, J. W. (1983). *Locational Change in the U.S. Semiconductor Industry,* unpublished Ph.D. dissertation (University of Washington).

Hayes, R. H. and Wheelwright, S. C. (1979a). 'Manufacturing process and product life cycles,' *Harvard Business Review,* January–February, 133–40.

Hayes, R. H. and Wheelwright, S. C. (1979b). 'The dynamics of process-product life cycles,' *Harvard Business Review,* March-April, 127–36.

Hill, T. and Utterback, J. M. (eds) (1979). *Technological Innovation for a Dynamic Economy* (New York, Pergamon).

Hoffman, W. D. (1976). 'Market structure and strategies of R & D behaviour in the data processing market—theoretical thoughts and empirical findings,' *Research Policy,* **5**, 334–53.

Horesh, R. and Kamin, J. Y. (1983). 'How the costs of technological innovation are distributed over time,' *Research Management,* March-April, 21–2.

James, F. and Struyk, R. (1975). *Intra-metropolitan Industrial Location: The Pattern and Process of Change* (Lexington, Ky.; Lexington Books).

Kamien, M. I. and Schwartz, N. L. (1982). *Market Structure and Innovation* (Cambridge, Cambridge University Press).

Kleinknecht, A. (1981). Observations on the Schumpeterian swarming of innovations, *Futures,* **13**, 293–307.

Kohn, M. and Scott, J. T. (1982). 'Scale economies in Research and

Development: the Schumpeterian Hypothesis,' *The Journal of Industrial Economics*, **30**, 239–49.

Kuznets, S. (1952). 'Comments on M. Abramovitz, Economies of Growth' in B. F. Haley (ed.) *A Survey of Contemporary Economics*, (Homewood, Ill., Irwin, for American Economic Association), Ch. II, p. 180.

Link, A. N. and Long, J. E. (1981). 'The simple economics of basic scientific research: a test of Nelson's Diversification Hypothesis,' *The Journal of Industrial Economics*, **30**, 105–9.

Malecki, E. J. (1979). 'Locational trends in R & D by large U.S. corporations, 1965–1977,' *Economic Geography*, **55**, 309–23.

Malecki, E. J. (1980a). 'Dimensions of R & D location in the United States,' *Research Policy*, **9**, 2–22.

Malecki, E. J. (1980b). 'Firm size, location and industrial R & D: A disaggregated analysis,' *Review of Business and Economic Research*, **16**, 29–42.

Malecki, E. J. (1980c). 'Corporate organization of R & D and the location of technological activities,' *Regional Studies*, **14**, 219–34.

Malecki, E. J. (1981). 'Science, technology, and regional development: review and prospects,' *Research Policy*, **10**, 312–34.

Malecki, E. J. (1982). 'Federal R & D spending in the United States of America: some impacts on metropolitan economies,' *Regional Studies*, **16**, 19–35.

Malecki, E. J. (1983). 'Technology and regional development: a survey,' *International Regional Science Review*, **8**, 89–126.

Mansfield, E. (1981). 'Composition of R & D expenditures: relationship to size of firm, concentration and innovative output,' *Review of Economics and Statistics*, **63**, 610–15.

Markusen, A. R. (1983). 'Profit cycles, oligopoly and regional transformation,' Working Paper No. 397 (University of California, Berkeley, Institute of Urban and Regional Development).

Martin, F., Swan, N., Banks, I., Barker, G., and Beandry, R. (1979). *The Interregional Diffusion of Innovations in Canada*, (Economic Council of Canada. Hull, Quebec, Supply and Services Canada).

Mensch, G. (1979). *Stalemate in Technology* (Cambridge, Mass., Balinger).

Meyer, J. W. (1978). 'Strategies for further research: varieties of environmental variation' in M. W. Mayer (ed.), *Environments and Organizations* (San Francisco, Jossey Bass).

Nabseth, L. and Ray, G. E. (1974). *The Diffusion of New Industrial Processes* (London, Cambridge University Press).

Nadiri, M. and Bitros, G. (1978). 'Research and development expenditures and labour productivity at the firm level' in J. Kendrick and B. Vaccara (eds), *New Developments in Productivity Measurement* (New York, National Bureau of Economic Research).

National Science Foundation (1983). *The Process of Technological Innovation: Reviewing the Literature* (Washington, DC).

Nelson, R. R. (1981). 'Research on productivity growth and productivity differences: dead ends and new departures,' *Journal of Economic Literature*, **19**, 1029–64.

Nelson, R. R. (1982). *Government and Technical Progress: A Cross Industry Analysis* (New York, Pergamon).

Nelson, R. R. and Winter, S. G. (1974). 'Neo-classical vs evolutionary theories of economic growth: critique and prospectus,' *Economic Journal*, 84, 886–905.

Nelson, R. R. and Winter, S. G. (1977). 'In search of useful theory of innovation,' *Research Policy*, 6, 36–76.

Nelson, R. R. and Winter, S. G. (1978). 'Forces generating and limiting concentration under Schumpeterian competition,' *Bell Journal of Economics*, 9, 524–48.

Nelson, R. R. and Winter, S. G. (1982). *An Evolutionary Theory of Economic Change* (Cambridge, Mass., The Belknap Press of Harvard University Press).

Nijkamp, P. (1982a). 'Long waves or catastrophes in regional development,' *Socio-Econ. Plan. Sci.*, 16, 261–72.

Nijkamp, P. (1982b). 'Technological change, policy response and spatial dynamics,' collaborative paper (Laxenburg, Austria, International Institute for Applied Systems Analysis).

Oakey, R. P. (1983). *Research and Development Cycles, Investment Cycles and Regional Growth in British and American Small High Technology Firms*, Discussion Paper No. 48 (University of Newcastle upon Tyne, Centre for Urban and Regional Development Studies).

Organization for Economic Co-operation and Development (1971). *The Conditions for Success in Technological Innovation* (Paris, OECD).

Pavitt, K. (1979). 'Technical innovation and industrial development,' *Futures*, December, 458–70.

Pennings, J. M. (1980). 'An ecological perspective on the creation of organizations' in J. R. Kimberley and R. H. Miles (eds), *The Organization Life Cycle* (San Francisco, Jossey-Bass).

Perroux, F. (1955). 'Note sur la notion du pôle de croissance', *Économie Appliquée*, 7, 307–20.

Piatier, A. (1981). 'Innovation, information and long-term growth,' *Futures*, October, 371–82.

Rees, J., Briggs, R. and Oakey, R. (1983). 'The adoption of new technology in the American machinery industry' Working Paper (University of Texas at Dallas).

Roberts, E. B. (1968). 'A basic study of innovators: how to keep and capitalize on their talents,' *Research Management*, July.

Roberts, E. B. (1980). 'Getting new ventures off the ground,' *Management Review*, June.

Roberts, E. B. and Wainer, H. A. (1971). 'Some characteristics of technical entrepreneurs,' *IEEE Transactions on Engineering Management*, 18, 100–109.

Rosenberg, N. (1982). *Inside the Black Box: Technology and Economics* (Cambridge, Cambridge University Press).

Rothwell, R. (1976). *Innovation in Textile Machinery: Some Significant Factors in Success and Failure*, SPRU, Occasional Paper Series, No. 2 (University of Sussex, Science Policy Research Unit).

Rothwell, R. (1977). 'The management of successful innovation in the firm,' *Proceedings,* December (Liverpool University, The School of Business Studies, Liverpool).

Rothwell, R. and Zegveld, W. (1982). *Innovation and the Small and Medium-Sized Firms: Their Role in Employment and in Economic Change* (London, Frances Pinter).

Saxenian, A. (1981). 'Silicon chips and spatial structure: the industrial basis of urbanization in Santa Clara County, California,' Working Paper No. 345. (University of California, Berkeley, Institute of Urban and Regional Development).

Schelling, T. S. (1978). *Micromotives and Macrobehavior* (New York, Norton).

Scherer, F. M. (1982). 'Inter-industry and technology flows and productivity growth,' *Review of Economics and Statistics,* 54, 627-34.

Schumpeter, J. (1950). *Capitalism, Socialism and Democracy* (New York, Harper and Row).

Schumpeter, J. (1971). 'The instability of capitalism' in N. Rosenberg (ed.), *The Economics of Technical Change* (Baltimore, Penguin Books).

Sirbu, M. A. Jr., Teitel, R., Yorsz, W., and Roberts, E. B. (1976). *The Formation of a Technology Oriented Complex: Lessons from North American and European Experience* (Cambridge, Centre for Policy Alternatives, MIT) (December 30).

Slater, M. (1980). 'The managerial limitations to the growth of firms,' *The Economic Journal,* 90, 520-8.

Solow, R. M. (1957). 'Technical change and the aggregate production function' *The Review of Economics and Statistics,* 39, 312-20.

Stead, H. (1976). 'The cost of technological innovation,' *Research Policy,* 5, 2-9.

Steed, G. P. F. (1982). *Threshold Firms,* Science Council of Canada, Background Study 48 (Hull, Quebec, Canadian Government Printing Centre).

Thomas, M. D. (1969). 'Regional economic growth: some conceptual aspects,' *Land Economics,* 45, 43-51.

Thomas, M. D. (1975a). 'Growth pole theory, technological change and regional economic growth,' *Papers of the Regional Science Association,* 34, 3-25.

Thomas, M. D. (1975b). 'Economic development and selected organizational and spatial perspectives of technological change,' *Économie Appliquée,* 29, 379-400.

Thomas, M. D. (1980). 'Explanatory frameworks for growth and change in multi-regional firms,' *Economic Geography,* 56, 1-17.

Thomas, M. D. (1981a). 'Growth, change, and the innovative firm,' *Geoforum,* 12, 1-17.

Thomas, M. D. (1981b). 'Perspectives on growth and change in the manufacturing sector' in J. Rees, G. F. D. Hewings, and H. A. Stafford (eds), *Industrial Location and Regional Systems* (Cambridge, Mass., Bergman/Ballinger).

Thomas, M. D. (1983). 'Economic growth and structural change: the role of technical innovation,' Discussion paper (Department of Geography, University of Washington), 47 pp.

Thomas, M. D. and LeHeron, R. B. (1975). 'Perspectives on technological change and the process of diffusion in the manufacturing sector,' *Economic Geography*, **51**, 231–51.

Thwaites, A. T., Oakey, R., and Nash, P. (1981). *Industrial and Regional Development*, Final Report (University of Newcastle upon Tyne, Centre for Urban and Regional Development Studies).

Thwaites, A. T., Edwards, A., and Gibbs, D. (1982). *The Interregional Diffusion of Production Innovations in Great Britain*, Final Report. (University of Newcastle upon Tyne, Centre for Urban and Regional Development Studies).

US Congress, Joint Economic Committee (1982). *Location of High Technology Firms and Regional Economic Development* (Washington, DC, US Government Printing Office).

Usher, A. P. (1971). 'Technical change and capital formation' in N. Rosenberg (ed.), *The Economics of Technological Change* (Baltimore, Penguin Books).

Utterback, J. M. and Abernathy, W. J. (1975). 'A dynamic model of process and product innovation,' *Omega*, **3**, 639–56.

Vernon, R. (1979). 'The product cycle in a new international environment,' *Oxford Bulletin of Economics and Statistics*, **41**, 225–67.

Vesper, K. H. and Albaum, G. (1979). 'The role of small business in research, development, technological change and innovations in region 10,' Working Paper (School of Business Administration, University of Washington).

Von Hippel, E. (1976). 'The dominant role of users in the scientific instruments innovation process,' *Research Policy*, **5**, 212–39.

Walker, W. B. (1979). *Industrial Innovation and International Trading Performance* (Greenwich, Conn., JAI Press).

Weintraub, E. (1979). *Micro-foundations* (Cambridge, Cambridge University Press).

3 Geography, Regions and Research and Development Activity: the Case of the United Kingdom

R. J. BUSWELL
R. P. EASTERBROOK
C. S. MORPHET
Newcastle upon Tyne Polytechnic

1. Introduction

Analysis of R & D establishment data has previously served to provide a basic description of the spatial distribution of research and development in the United Kingdom (Buswell and Lewis, 1970). In the first part of this chapter we provide an update of this analysis on the basis of information published in 1983. The strong locational bias towards the South East, and the consequent under-representation of the peripheral regions, remains visible. We introduce a number of caveats to the interpretation of such data, and continue in the second part of the chapter to suggest a possible framework for the further analysis of R & D location. This framework reflects recent developments in industrial geography which impose considerations of industrial segmentation (e.g. Taylor and Thrift, 1983) and economic long-waves (e.g. Rothwell, 1982), alongside some of the industrial geographers' classical concerns with communication, agglomeration and locational preference. We are particularly concerned to establish that variations between firms in their R & D expenditure will reflect real differences in the nature of these firms, and thus in the demand for R & D.

Three main factors are identified which we argue underlie the relevant variation between firms, and which in turn underlie the geography of R & D:

(i) industrial sector, where sectorial differentiation is not necessarily bound by SIC or MLH categories but is primarily based on the position of a productive sector in relation to the Kondratiev cycle of growth, maturity and decline;
(ii) the relationship of the industrial sector to the defence industry;
(iii) the 'technological role' of a firm within the industrial system. In defining this we draw upon ideas of segmentation as described by Taylor and Thrift (1983), and the associated ideas of innovation strategy as described by Freeman (1982).

Our argument suggests a three-dimensional classification of business organizations. These three dimensions are shown in Fig. 3.1. It is suggested that along each axis, proximity to the origin produces a tendency towards a geographical core location. In this sense the space described

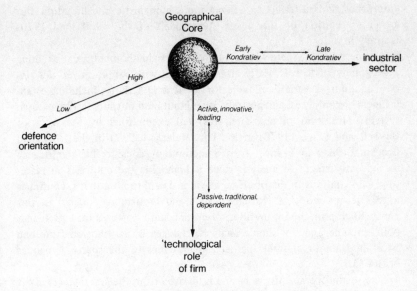

Figure 3.1 Geographical peripherality in R & D?

by Fig. 3.1 can be seen to relate to geographical space. As Taylor and Thrift (1983) have suggested, a multidimensional approach to the segmentation of business organizations is made difficult because the dimensions selected will not be truly independent. We admit that covariation between our three variables will exist, and return to this issue in our conclusion.

2. Theoretical approaches to the geography of R & D

Until recently only a limited number of perspectives on the geography of research and development activity could be identified, nearly all of which were products of their time, parts of the prevailing paradigms of geography and the social sciences. Although it may now prove possible to provide a more radical interpretation of the present-day spatial patterns of R & D, that view too is a reflection of the contemporary concern of geographers about the role of capital in locational decision-making processes. However we interpret the map of R & D before us, it is difficult to do so without being influenced by the ways in which both geographers and students of the social relations of science and technology formulate and communicate their ideas. Obviously, for such an overtly *economic* activity, and one that is so concentrated in large-scale capitalist or state enterprises, it is difficult to offer a more humanistic interpretation, but a reinterpretation of the positivistic models of the 1960s and 1970s is surely in order.

Formerly, then, there were a number of starting-points for constructing a geography of R & D. Early attempts began with simple *spatial description*, mapping a variety of indicators of R & D activity, including establishments, employees, and possibly other inputs and outputs to the research system. This kind of approach was well exemplified by the work of Buswell and Lewis (1970) in the UK, Malecki (1979) in the USA, and Brocard (1981) in France. It is a method of considerable significance in the early stages of analysis since it highlights the contrasts in place-to-place variations in distribution, and it is useful to note that it continues in the recent work of Howells (1984) and is furthered in part of this paper. Such a method is useful in describing static patterns for a particular point in time and, by simple analysis, changes in distribution from one point in time to another, perhaps giving clues to the spatial dynamics of R & D.

The second perspective was, and is, derived from the *locational analysis* school, which treated the distribution alluded to above as point patterns, or possibly areal patterns, which could be interpreted via our understanding of locational decision-making and model-building. Thus description gave way to theory and the model design to empirical testing. The elements thought to influence locational decisions could be isolated, weighted and calibrated, leading to testing in the field or from published data sources. This approach has been described elsewhere (see Buswell, 1972, for example), and it is clear that whether the single laboratory or punctiform approach was adopted or whether some kind of ecology/minimum energy model based on linkage structures was used, firms were seen to be maximizing output gains, minimizing costs, or adopting some kind of optimizing activity. The behavioural revision of such models added little other than to show that firms were obviously behaving suboptimally in some way, and therefore, that locational decisions were not being made on the basis of positivist economic ideas. Behavioural analysis at first gave very little insight into the spatial decision-making process in R & D. It was apparent, however, that location was a product of the firm's behaviour somewhere along the line, and this stimulated the application of ideas on corporate behaviour rather than entrepreneurial behaviour. The 'theory of the firm' suggested to geographers that R & D might be treated by corporations in a rather more complex way than was appreciated, and some benefit was derived by including R & D in the high-order category of firms' functions and drawing insights from the processes thought to be at work in, for example, a headquarters location. It was not then a large step to a more critical analysis that envisaged corporate capital managing geographical space in a particular way, perhaps as part of Stuart Holland's *meso-economy* (Holland, 1976).

The third approach might be described as the *regional analysis* approach, which is predicated on two complementary notions: (a) that interregional variation in the geographical distribution and location of R & D is in some way 'significant', reminiscent perhaps of the 'equity vs. efficiency' arguments that so often underlie regional economics in mixed economies; and (b) that such 'significance' could be altered by government intervention to bring about a different (and usually supposedly better) set of distributions. Thus in many countries the geographical study of R & D has become enveloped in regional policy with a belief that enhancement of R & D in structurally or locationally 'backward' regions will lead to greater levels of technological advancement (see Buswell, 1983, for example), in turn leading especially to new product innovation (Rothwell, 1982) and increases in regional income (export-based) and/or more employment.

These three perspectives, readily detectable in the literature of the geography of R & D, are derived very much from mainstream ideas in economic geography. Malecki (1983), in a recent review of technology and regional development, has concluded that the relationship between the two is more complex than neo-classical analysis would suggest. He writes (p. 99) that

> simple agglomeration-related patterns of innovation are difficult to integrate with the activities of corporations regarding product mix, employment and growth or decline. The disaggregation from a region with a single capital/labour production function to one with many firms of all sizes in a varying mix of industries represents too great a challenge at present.

But after twenty years of such study, what more do we now know, or in what way has such knowledge really affected the regional imbalance described therein? Have *geographers* really read widely enough into the literature of R & D itself—that is, R & D as a social, economic and political activity rather than a spatial one—and to what extent have they really absorbed the political economists' teaching on the organization of capital? Could it be likely that more progress in our understanding of the geography of R & D might be made by probing the way in which decisions are made by large corporations and by the state, through a more detailed examination of the organizational framework for such decisions?

Assuming the dualism of private and public sectors, four explanatory frameworks might be identified:

(a) national strategic considerations, largely determined by the state and related primarily to military research and the defence industry;

(b) corporate strategies in the private sector, often related to (a) above, since about 30 per cent of manufacturing industry's R & D expenditure is devoted to the country's defence industry. Freeman's identification of a range of strategies by private firms may suggest ways in which geographical space might be utilized to further their particular aims (Freeman, 1982). Whether the organizational structures identified by Malecki (1980) lead to the clear dichotomy of 'general' versus 'specific' locations for R & D is open to debate;

(c) against this background of state or corporate involvement, it is evident that in mixed economies, at least in principle, government policy strategies might influence locational decisions as, for example, was attempted in France. This might be done as part of regional policy or as part of a wider strategy for industrial development (see, for example, Brocard, 1981);

(d) locality strategies might be mounted whereby certain cities or sub-regions might seek to attract R & D activity as a means of diversifying local economies. The current popularity of the development of science parks seems to underline this kind of strategy, and where the interests of capital and local authorities coincide then we might expect new locations to be effected. Certainly Peterlee New Town in North East England developed this strategy in the early 1970s (without success) and the new booster towns of the American South West are actively hunting for R & D operations, as is nearly every British provincial city with a university.

Of these four frameworks, only the first two would seem to be of major significance in shaping the geography of technological change. The sheer size of the defence industry, together with the concentration that has gone on in British manufacturing industry since the early 1970s, has resulted in a considerable shift in the centre of gravity of British industry which the final two approaches—essentially policy-orientated interventionism—can scarcely hope to modify. This contentious point of view has to be argued against a background of academic research in geography which has seen British manufacturing becoming more widespread either interregionally (Keeble, 1976) or intraregionally (for example, Fothergill and Gudgin, 1982; Keeble, 1980). However, much of this research was based on *employment data*, and it might be argued that in the light of (a) the massive recession in manufacturing employment since 1979 in which over 14 per cent of all such jobs have disappeared, and (b) what is now known about the corporate strategies of firms (for example, Massey and Meegan, 1979), the most meaningful indicators as to present and future developments might lie within *establishment-based data*. This could prove to be especially true for establishments performing different functions.

Although some have argued for a product cycle model to explain spatial hierarchies of establishments (for example, Massey's statement that the spatial separation of research and development work from production is a reinforcement of the 'spatial concentration of control functions' (Massey, 1978, 49)), it is evident that corporate decision-making produces more complex patterns than this product cycle model would suggest. The emerging detailed empirical work of Townsend and Peck (1984) reveals the varieties of strategies that firms adopt with reference to plant closure and redundancy, and therefore perhaps by implication to R & D decisions. However, at a simple level the dichotomies identified by Massey (1978) (capital and labour; control and production; skilled and less skilled labour) remain attractive starting-points in any analysis, and ones which substantially inform our own provisional attempt to analyse the 'structure' of research and development. To this we turn, after we have presented an analysis of the most recently available data on the location of R & D establishments, and after we have commented on its significance.

3. The analysis of R & D establishment data

3.1 *The evidence*

The work of Buswell and Lewis (1970) gave an early indication as to the geographical distribution of R & D in the United Kingdom, suggesting in particular that directory-based information could reveal significant contrasts in the locations of R & D units of industrial firms. Table 3.1 shows the relative and absolute distributions of these units by region for 1968, drawn from this study. These data are compared with analyses of similar directory-based data for 1980 (Department of Trade and Industry, 1983) and 1983 (from Williams, 1983). The dates refer to different editions of similar directories, and in fact entries are compiled over a slightly longer period of time. Table 3.1 refers only to private industry, and does not include public corporations or independent private and sponsored laboratories. It displays a remarkable similarity in the relative regional distributions over a fifteen-year interval. The major change in

Table 3.1 Distributions of industrial firms' R & D units by region, 1968, 1980 and 1983

Region	1968 No. of units	1968 % of total	1980 No. of units	1980 % of total	1983 No. of units	1983 % of total
Scotland	41	5.7	16	3.8	14	4.5
Wales	11	1.5	9	2.2	6	1.9
Northern Ireland	7	1.0	–	–	2	0.6
Northern	35	4.9	21	5.0	18	5.8
North West	79	11.0	48	11.5	39	12.7
Yorkshire and Humberside	46	6.4	26	6.2	19	6.2
UK	720	100	417	100	308	100
West Midlands	82	11.4	37	8.9	27	8.8
East Midlands	40	5.5	23	5.5	15	4.9
East Anglia	14	1.9	11	2.6	8	2.6
South East	325	45.2	196	47.0	140	45.5
South West	40	5.5	30	7.2	20	6.5

Note: 1980 figures refer to GB only, not UK
Sources: Buswell and Lewis, 1970, p. 302; Department of Trade and Industry, 1983, p. 57; Williams, 1983.

fact is the dramatic decline in the absolute numbers of these units, which will be considered later. However, there is a need for careful interpretation of this establishment-based data in each case. The basis for including establishments in the earlier studies is not known, but the analysis of the 1983 data is based on the *locations* of entries in the directory where R & D is being carried out. This is because not all entries relate to research locations, and in some cases each location has more than one entry. This means that in cities, whilst the same or an adjacent address does not merit a further R & D location being recorded, a location elsewhere in the city does. Since the entries are nominated by the firms themselves, in response to a questionnaire, the degree of coverage of research or development activities may vary considerably. Thus any one firm may include a wide range of development locations at divisional production sites, for example, whilst another may only include the location of its major corporate R & D site. This degree of variation may become crucial if a regionally-based firm is overrepresented, given the very low numbers of locations in some cases. It clearly becomes difficult to talk in general terms about 'factors of location' of R & D given that, at the very least, we do not know the relative contributions of research as against development work in these distributions. In any case, the actual *amounts* of R & D carried out, whether measured as expenditure, employment, or any other input or output factor, also vary widely between R & D locations. Expenditure, for example, ranges from the minimum of £30,000 per year needed for inclusion in the 1983 directory, to the very large sums spent at corporate research centres such as the British Petroleum group research centre at Sunbury-on-Thames, which in 1982 accounted for about £50m (*Financial Times*, 2 February 1982). For all these reasons we cannot be confident that the distribution of these units really reflects the national distribution of private industry R & D, although on the other hand there is no known reason for any spatial bias in the actual compilation of the directories.

For the first time it is possible to see the actual point distribution of these units, rather than the aggregate regional picture shown in Table 3.1, and this is presented for 1983 in Figure 3.2. Given the caveats to the interpretation of the data, this map does, none the less, demonstrate that there is considerable variation in the degree of spatial concentration below the regional scale. The importance of the band of units in West London and neighbouring parts of the Home Counties is particularly striking, as is that of the new town locations. Crawley, Bracknell, Welwyn Garden City and, in particular, Harlow, are all represented, a fact which Buswell and Lewis (1970) pointed out in the case of the 1968 data. In general terms, the relative importance of locations at the subregional

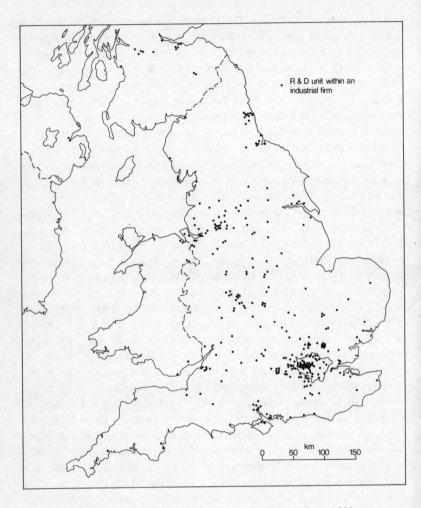

Figure 3.2 Locations of R & D units of industrial firms, 1983

scale merits further study; however, it would be wrong to speculate at this stage about how 'urban' or 'rural' these locations are, or how residentially attractive, or how related they are to major universities, without further evidence. For example, whilst Howells (1984, 18) has suggested that the location of employment in R & D services in Cambridgeshire was 'no doubt due to the presence of Cambridge University', it is not clear in the present study as to the locational influence of, for example, Harlow Technical College.

There are other crude measures of relative over- or underrepresentation of R & D units by region which might be considered. Table 3.2 indicates that while the South-East still dominates the relative distribution, the Northern and South-West regions do seem to have R & D unit locations on a par with their shares of total manufacturing employment in the UK. This analysis suffers from the fact that the use of total manufacturing employment data ignores the possibility of regional variations in industries carrying out most R & D, notably electronics and aerospace. Some evidence on this latter score is provided by Gudgin, Crum and Bailey's (1979) account of the location of non-production employment in manufacturing which, although not referring to R & D employment specifically, suggested that regional variations in scientific and technical personnel owed little to variations in 'industrial structure' in this sense.

One way of gaining a more accurate picture of R & D in *aggregate* is to turn our attention to the few large enterprises whose activities dominate the national industrial R & D effort. Some idea of the magnitudes of individual enterprises' R & D programmes, by expenditure, can be gained from the *World Directory of Multinational Enterprises*. For example, in 1981, ICI spent about £229m on R & D, £170m of which went to the UK; Plessey, about £135m; Lucas Industries, £65; Glaxo Holdings, £45m;

Table 3.2 Regional shares of R & D units, 1983, compared with regional shares of UK manufacturing employment, 1981

Region	Ratio of percentage of R & D units, 1983 to percentage of manufacturing employment, 1981
Scotland	0.60
Wales	0.46
Northern Ireland	0.30
Northern	1.00
North West	0.93
Yorkshire and Humberside	0.64
West Midlands	0.68
East Midlands	0.57
East Anglia	0.87
South East	1.75
South West	1.00

Sources: Williams, 1983; Census of Production, 1981.

and so on. Whilst all these firms have entries in the 1983 directory of R & D establishments, such very large amounts of expenditure cannot be accounted for in total by the listed R & D units, even taking into account that not all the expenditure will occur in the UK. These figures do not necessarily correspond with the 'Frascati' definitions of R & D used in official UK statistics, but in general it is clear that large enterprises are very important. In the size analysis of UK private industry enterprises in *Business Monitor* MO14, referring to 1978, fifty-three enterprises (6.4% of the analysis), each with a total employment of 20,000 or more, accounted for 66 per cent of R & D employment and 66 per cent of gross R & D expenditure. For the purposes of *Business Monitor*, 'private industry' includes some companies in which the government has a major shareholding, but these are compatible with the 'industrial firms' whose R & D units are analysed in Fig. 3.2. The degree of dominance of a few large enterprises is shown even more clearly when the largest R & D programmes are considered, rather than the largest numbers of total employees. In 1978, the five enterprises (0.6 per cent) with the largest R & D programme expenditures accounted, alone, for approximately 41 per cent of gross expenditure and 40 per cent of R & D employment. This is not to suggest that, in general, very large enterprises spend disproportionately more on R & D. However, it is worth emphasizing the point that at the national scale the geography of R & D is intrinsically related to the geography of relatively very few enterprises, and that this relationship is even more marked in the case of government-funded research carried out in private industry.

3.2 *Research and development and the requirements of the firm: the diversity of approaches*

In order to consider the possibility that there has been a major change in the numbers of R & D units, if not the locations, it is necessary to examine the dynamics of R & D. Here, we might usefully consider trends in the spatial separation of functions within the multiregional enterprise, taking into account the changes in industrial concentration since the mid-1960s (Watts, 1980), and, similarly, consider the changes in production accompanying the changes in the level and nature of manufacturing activities over this time. It is unrealistic to separate these factors since, as Massey and Meegan (1979, 161) have pointed out, 'locational behaviour of firms is a product of macro-structural forces, rather than of a locational decision which can be understood at the level of the firm alone', and their case studies of employment change in a spread of industrial sectors (Massey and Meegan, 1979, 1982) illustrate this well. Their

1979 study concentrated on the restructuring of production between different firms in the electrical and electronic engineering sectors aided by the IRC as a response to various competitive pressures. In some cases there was a specific R & D element in this restructuring, where the nature of the competitive threat stimulating restructuring was at least partly technological. Here there was a need to increase the absolute level of R & D spending within one enterprise, through merger, to enable a threshold expenditure for international competition to be reached. Where this type of restructuring occurred, R & D facilities were concentrated into a smaller number of larger groupings, which had direct spatial effects. For example, in the English Electric/Elliott-Automation merger of 1967, Elliott's R & D laboratory at Glenrothes was closed, work being concentrated at Chelmsford-Great Baddow (now GEC–Marconi Electronics). In the formation of ICL, under the Industrial Expansion Act (1968), again involving considerable restructuring, two development and design departments were closed and the work concentrated elsewhere.

Massey and Meegan suggest that these changes portend hierarchies of location within the multiregional firm, although these examples give no clear regional distribution of closure and concentration. They are at pains to point out that such locational effects are a result of the stimulus to the merger, rather than the spatial effects of organizational integration, or other long-term changes. Given this, we might usefully extend the analysis to look at the reorganization of production within and between firms in response to corporate strategy which itself interacts with macro-scale processes. Here, too, we might expect a variety of direct technological strategies and other strategies with organizational or long-term effects on R & D. For example, in the case of a rationalizing sector, we might expect large enterprises to concentrate production in the most productive plants, which will inevitably have an effect on the location of associated development work, in this case probably process-orientated. What we do not yet know is the frequency, extent or duration of these changes, which can only be fully explored at the level of the case study; in particular, we would need much further work to establish whether there actually has been a concentration of R & D units as suggested by the reduction in the number of establishments in the two recent R & D directories.

4. The geography of R & D: a new interpretation

4.1 *The importance of industrial sector*

The role and nature of R & D (and thus its locational requirements) will vary between industrial sectors. We suggest, further, that the variation will be unlikely to respect standard sectoral boundaries or even those of the Minimum List Headings of the Standard Industrial Classification. Distinctions have often been drawn in the past between science-based industries and those in which the technology is more traditional or mature. It has been argued (see, for example, Green and Morphet, 1977) that such distinctions can help to explain variations in research intensity between industries. It will be suggested here that a comparable set of distinctions will help to explain variations in the spatial requirements for R & D.

The basis for the sectoral differentiation that we use is that afforded by the Kondratiev theory (Kondratiev, 1935) used here as a descriptive account of the growth and maturation of various industrial sectors in the UK. Rothwell (1982) has outlined the post-war evolution of industry in the UK in terms of Kondratiev cycles, and has discussed the role of technology in Kondratiev change. It has been argued that the 1980s will see the onset of a new Kondratiev cycle based on the so-called 'new technologies', which include microelectronic and information systems as well as biotechnologies. We now have in the UK the industries of the 4th (post-war) Kondratiev cycle, which include electronics, petrochemicals and plastics, and pharmaceuticals, as well as industries remaining from previous Kondratiev cycles. These include electric power and motor vehicles from the 3rd Kondratiev, steel and heavy engineering from the (railway-dominated) 2nd Kondratiev, along with the remains of the driving industries of the 1st Kondratiev, largely the textile industries launched with the aid of steam power during the industrial revolution.

Rothwell and Zegveld (1979) provide a schematic representation of the technical and economic features of the introduction of new technologies according to the Kondratiev concept. The features described in Table 3.3 are adapted from this account.

While Rothwell (1982) and others have described the spatial implications for the location of *production* in various stages of the Kondratiev, the present intention is to explore further the implications of its several stages for the nature, and consequently the location, of *R & D*. Different sectors of the economy, representing industries of different Kondratievs at different stages of evolution, will thus be seen to have different locational requirements for R & D. For the present purpose we have chosen

Table 3.3 Technical and economic features of the Kondratiev cycle*

Previous Kondratiev
   Basic science coupled to technical exploitation. Key patents. Many
   prototypes. Rapid design changes. No standardization. Some disasters.
   Production one-off moving to small batch, closely linked to R & D
   and design. Consumer resistance and ignorance. Employment for
   skilled labour, engineers and technicians.

Upswing of main Kondratiev
   Intensive applied R & D for new products and applications, and for
   trouble shooting. Movement to larger batch or flow processes, economies of scale emerging. Technological competition for design and
   performance, falling prices, imitative adoption by consumers. Demand
   for skilled labour.

Downswing of main Kondratiev
   Intensive applied R & D but shifting to cost-saving process innovation.
   Emphasis on standardization, economies of scale, price competition.
   Strong pressure for economies of scale through export markets. Skilled
   labour supply catching up with demand.

Subsequent Kondratiev
   Innovation limited to routine design or incremental technological
   change. Output growth achieved through productivity increase. Routine
   mass production; oligopoly or monopoly arising from bankruptcies or
   mergers.

---

* Adapted from Rothwell and Zegveld, 1979.

to consider two main sectoral types, the first identified with the upswing
stage of a Kondratiev, although possibly occurring even prior to this;
and the second with the downswing, or the later mature phase, in which
dynamic growth resides in the industries of a subsequent Kondratiev.
Since these two sectors are described in terms which assume a market
economy, we add a third sector for consideration as a further dimension
(see Fig. 3.1): the huge and highly research-intensive defence sector within
which R & D location can be considered in its own unique terms.

(a) *Sectoral classification on the basis of Kondratiev theory*
   (i) Industries at the emergence of a Kondratiev cycle. As Table 3.3
suggests, the early period of a Kondratiev cycle is marked by an emphasis
on innovation in new industries, and on a research base for these industries
which must be responsive both to changing perceptions of (the new)
markets and to changes in world science and technology. Such industries

are plainly 'research-based' in a profound sense, and many of the traditional locational determinants of R & D might be applied to them. Thus agglomeration advantages, proximity to major centres of world science and technology, and a close coupling with market research, will all indicate a location at the national geographical core.

This view can be associated with, and supported by, the analysis of the role of science and technology in restructuring offered by Blackburn, Green and Liff (1982). This serves (indirectly) to underline the pressures for a core-region location for industries based on new technologies and, in so doing, offers two new and interesting insights: firstly, that the restructuring of production nearly always has to be preceded by the restructuring of consumption, and secondly, that the radical critique of the role of science and technology offered by Mandel (1975), for example, could now be applied to the service industries, which are today the dominant employers in most developed societies with approximately 60 per cent of total employment.

What are the implications of the first observation for the geography of R & D? The restructuring of consumption would appear increasingly to determine research strategies. This would lead to a continuing refinement of existing product lines interspersed with new products generated by market research. R & D would therefore need to be in much greater sympathy with markets, which in socioeconomic and spatial terms would usually mean the core areas of states from whence demand is traditionally led and where it is quantitatively concentrated. Thus the high density of the R & D functions of the new industries in the South East region of the United Kingdom may not be a product of energy-minimizing or linkage-enhancing, location-modelling decisions but, rather, a response to the need to be at the centre of market activity.

The second observation suggests that, if an earlier generation of more radical thinkers saw the process of technological change as a means by which capital sought to deskill or replace labour, then it might now be possible to argue that the service sector shows evidence of similar changes. In manufacturing, R & D might be seen as acting in the interests of capital to reduce costs (process R & D) or to diversify production and stimulate new demand (product R & D). In the service industries labour is often perceived by capital as being underproductive and therefore the new 'labour displacing information technologies form a technological arsenal to attack the growth of unproductive labour' (Blackburn, Green and Liff, 1982, 21). Thus, if the use values of consumers have to be satisfied as well as the exchange value of labour in the manufacturing sector, then in the service industries, too, efforts will be made by capital to alter the relations between provision and consumption by transforming the labour process that provides them. This might be achieved by introducing new technologies or, as Gershuny (1978) has suggested, by making many services amenable to *self*-service. This in turn will require a larger investment in domestic technologies to replace public and private services (for example, the potential effects of the introduction of home computers) and in the education service to teach individual consumers to use such technologies, eventually leading to an increasingly privatized provision of many services in the public sector.

(ii) Industries of a previous Kondratiev cycle. Industries such as steel, heavy engineering and textiles are the products of a previous Kondratiev cycle. While they may or may not be in decline, they are no longer part of the growth processes which drive the economy. As Table 3.3 shows, they are characterized by a *production* orientation. Low-cost mass production, rather than product innovation or technical product differentiation, is the order of the day. Innovation will be aimed at productivity increase (process innovation) rather than product innovation, and such product innovation as exists will be incremental. It is such industries, with a production orientation, that will have exploited regional incentives or will otherwise have found advantage (see, for example, Massey, 1979, 1983) in peripheral locations. Even where corporate structures tend to centralize R & D, branch plants at peripheral locations will still attract the more routine aspects of development work, perhaps work towards the boundary of R & D where development shades into maintenance, and where day-to-day contact with the production process is necessary to develop improved performance from automated or semi-automated fabrication, assembly or process machinery.

(b) *Sectoral classification with reference to defence orientation*. A further observation stems from the increased and increasing role of the military and defence industries within most sectors of contemporary industrial activity. As a recent Defence Committee report noted, 'the defence industrial base is a significant factor in national patterns of employment, the development and use of technology and the performance of the economy as a whole' (Defence Committee, 1982, para. 119). In 1980–1, the Ministry of Defence placed some 30,000 contracts to a total value of approximately £4.0bn (Defence Committee, 1982). Out of a total defence budget of £12.3bn for 1981–2, £5.4bn, or 44 per cent was for defence equipment, 'an enormous programme of direct purchase of goods and equipment from industry unparalleled in any other government department'

(Defence Committee, 1982, para. 2). For 1982-3, the value of defence industry exports was estimated at approximately £1.8bn (NEDO, 1983). This military demand supports 220,000 jobs in direct employment in defence industries and, indirectly, about the same number in industry as a whole. Between 1976-7 and 1981-2, the Ministry of Defence's R & D budget rose 135 per cent; approximately 26 per cent of its procurement spending is now devoted to R & D, of which £300m is spent on research and £1,540m on development work (NEDO, 1983). Defence R & D accounted for 53 per cent of government expenditure on R & D in 1980-1, and contracts from MoD represented approximately 31 per cent of the R & D undertaken in private industry and public corporations in the same year. In a memorandum from the Electronic Engineering Association to the Defence Committee (Defence Committee, 1982), it was shown that in this one sector of industry—electronics—Ministry of Defence expenditure on radio, radar and electronic capital goods in 1979-80 was £540m, with £84m spent on components. In addition, the value of electronics in defence exports was estimated at £300m-£400m, so that the total value of the defence industry to electronics was in excess of £1.0bn per annum. Some industrial groups are totally dominated by the military economy: British Aerospace, for example, now has only 20 per cent of its activity in the civil sector, and it is hardly surprising that other private sector corporations are anxious to buy into such companies when they are privatized. A recent market research publication observed that 'it is now the case that MoD buys about 20% of the output of the British electronics industry, about 33% of the shipbuilding industry and about half of the aerospace industry' (Jordans 1981, iv).

Translating the evolution of what Mandel (1975) calls the 'permanent arms economy' into the language of the geography of R & D is not an easy task, if only because of the secrecy that surrounds so many of the contractual details of military spending. Few would deny the key role that innovation for military purposes has played in contemporary industry; even if it is accepted that military demand rarely leads to invention, it does provide a rich environment for *innovation*, especially under active wartime conditions. In recent years, the British government, whilst stimulating the growth of military spending, has become alarmed at the effect of the size and rate of this growth (over 3 per cent per annum), and, in an attempt to contain such public spending and to obtain 'value for money', has sought a greater co-operation between the military and civil sectors so that the products of the former can be adopted and adapted by the latter for commercial development. The Strathcona Report (Ministry of Defence, 1980) has already suggested that some of the MoD's R & D should be devolved to private industry. In addition, the thrust of the recent Defence Committee (1982) report is a wish to see the transfer of more of the 'risk' of military equipment innovation away from the public sector, where the report considers the risks cushioned, to the private sector, and to see a rise in the exports of arms and military equipment in order to offset domestic costs. In 1980, Britain ranked fifth in terms of defence equipment exports, with 3.7 per cent of the world's total (compared with the USA, 43.3 per cent; USSR, 27.4 per cent; France, 10.8 per cent; and Italy, 4.0 per cent), and was characterized by a growth rate over the twenty years since 1960 of 29 per cent, compared with the USA's 700 per cent and France's 2,400 per cent (Jordans, 1981, vi). Sir Ieuan Maddock has recently cast considerable doubt over the ability of the large defence-orientated corporations to develop into civilian markets (NEDO, 1983). However, it is conventionally suggested that this can only be achieved by allowing private industry to design and manufacture defence equipment for as wide a range of markets as possible, and not specifically for British military consumption—hints here of Blackburn, Green and Liff's consumption restructuring, perhaps. It is clear that the production of complex defence systems "involves a number of sub-system specialists whose activities need to be co-ordinated and integrated to meet an overall performance specification efficiently, economically and to time' (Defence Committee, 1982, para. 85). The responsibility for the management of total production is increasingly to be passed to a prime contractor, which might suggest to some that further corporate acquisition activity will be required, or possibly better and closer spatial liaison between linked firms, or both.

In geographical terms, then, it could prove valuable to study the location and distribution of the defence and defence-related industries in both the private (see Fig. 3.3) and public sectors, and, especially, the spatial patterns of R & D that support, and are supported by, the military economy. Preliminary analysis suggests a key role for the South East region of the UK, where the Ministry of Defence locates its procurement office and officers, where nearly all of the government's defence research is located (see Fig. 3.4), and where some of the major testing grounds for military equipment are found (important for export sales with access through the South East's international airports). If the speed with which military innovations are to be transformed for private consumption is to be increased, then clearly the South East region offers the same kind of environment for this kind of activity as it does for conventional new product marketing. Maddock believes that the smaller technology-based entrepreneurial companies may be able to break 'the high degree of

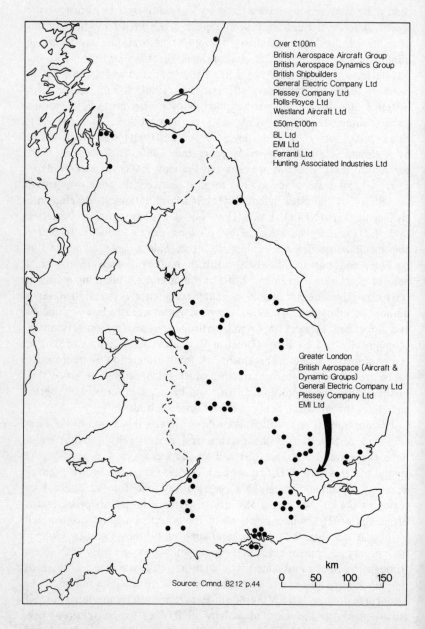

Over £100m

British Aerospace Aircraft Group
British Aerospace Dynamics Group
British Shipbuilders
General Electric Company Ltd
Plessey Company Ltd
Rolls-Royce Ltd
Westland Aircraft Ltd

£50m–£100m

BL Ltd
EMI Ltd
Ferranti Ltd
Hunting Associated Industries Ltd

Greater London

British Aerospace (Aircraft &
Dynamic Groups)
General Electric Company Ltd
Plessey Company Ltd
EMI Ltd

km

0   50   100   150

Source: Cmnd. 8212 p.44

Figure 3.3 Sites of UK-based MoD contractors paid £50m or more by MoD for equipment, 1979–80

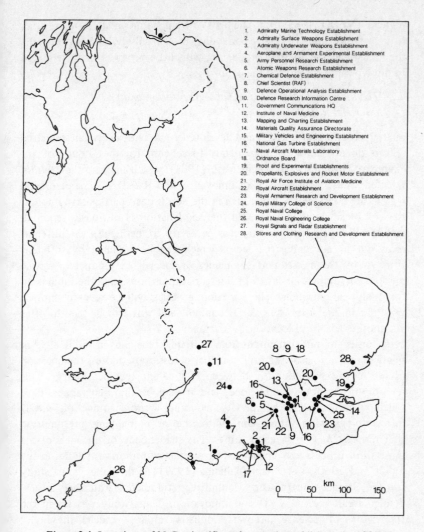

1.   Admiralty Marine Technology Establishment
2.   Admiralty Surface Weapons Establishment
3.   Admiralty Underwater Weapons Establishment
4.   Aeroplane and Armament Experimental Establishment
5.   Army Personnel Research Establishment
6.   Atomic Weapons Research Establishment
7.   Chemical Defence Establishment
8.   Chief Scientist (RAF)
9.   Defence Operational Analysis Establishment
10. Defence Research Information Centre
11. Government Communications HQ
12. Institute of Naval Medicine
13. Mapping and Charting Establishment
14. Materials Quality Assurance Directorate
15. Military Vehicles and Engineering Establishment
16. National Gas Turbine Establishment
17. Naval Aircraft Materials Laboratory
18. Ordnance Board
19. Proof and Experimental Establishments
20. Propellants, Explosives and Rocket Motor Establishment
21. Royal Air Force Institute of Aviation Medicine
22. Royal Aircraft Establishment
23. Royal Armament Research and Development Establishment
24. Royal Military College of Science
25. Royal Naval College
26. Royal Naval Engineering College
27. Royal Signals and Radar Establishment
28. Stores and Clothing Research and Development Establishment

Figure 3.4 Locations of MoD scientific and research establishments, 1983

vertical integration in the larger groups' (NEDO, 1983, 16). An analysis of the regional implications of this view would surely suggest that in the UK only the South East and possibly a limited number of other subregions would have the capacity to reform the British electronics industry in the way he suggests. In terms of this industry Maddock observes that 'the case of the civil platform has been subsiding, leaving the defence "peaks" standing even higher relative to the national Electronic Engineering

plateau' (NEDO, 1983, 8); in the light of what is known about the restructuring of this industry, the regional analyst can only point to the benefits
that might accrue to the core region, with little prospect of their spread
to more peripheral parts of the country.

4.2 *'Technological role'—industrial segmentation and industrial R & D*

Spatial variations in research and development activities between industrial
sectors do not allow us immediately to account for the findings of, for
example, Thwaites, Oakey and Nash (1981), which indicate that, *within*
selected industries, the levels of innovation and R & D in some peripheral
regions are considerably lower than in the South East, particularly amongst
the set of single-plant firms. Further explanation is obviously required,
and it will be argued here that it is necessary to distinguish the different
roles within any industrial sector which can be assumed by different
firms or by the constituent companies or specialized product or service
divisons or subdivisons into which larger corporations might be subdivided
for analytical purposes. The variation in roles will be seen to imply a
variation in the sorts of R & D required, and this will be seen to offer
yet another source of locational variation.

In order to understand the role that might be played by R & D in
relation to the role of the firm, it is first necessary to address the question
of the narrower relationship between R & D and technical change. This
relationship is not a simple one, and it may be that geographers have
failed to appreciate its complexity, assuming a simple model of 'R & D
in/technological-innovation out' with the accompanying presumption
that more R & D is 'a good thing'. The inadequacy of this view of the
innovation process was challenged in a series of important studies in the
1950s and 1960s. Carter and Williams (1957), in their pioneering study
for the Science and Industry Committee, straddled a watershed in academic research on R & D and innovation. Their study was set in the
framework of 'science discovers, industry applies'. Indeed the book,
entitled *Industry and Technical Progress*, was subtitled 'Factors Governing
the Speed of Application of Science'. And yet the authors were plainly
aware of the inadequacy of the 'linear-sequential' view or, as they put
it (p. 54), the

> misconception . . . that research is naturally a left-to-right process—
> that is that fundamental research produces something that is communi-
> municated to the industrial scientist, who performs some applied research
> and communicates the results to someone else who takes matters a step
> further. . .

and, as Williams (1967, 57) later explained, the erroneous but

> widespread belief that the speed of industrial innovation is a simple function of research. Think how often we are told that we ought to be doing more research to ensure a higher rate of innovation (or growth).

Carter and Williams's (1957) crucial chapter in this respect was entitled 'Pushes and Pulls', and it sums up very neatly the key questions in the R & D innovation relationship. It begins (p. 108):

> There can be no scientific progress without scientists, but it is not at all certain that, if more scientists are trained, unprogressive firms will leap into progressiveness. The various necessary conditions of progress—trained men, money, receptive managements, favourable markets—are essential to each other like the wheels of a clock. But what winds the clock or makes it tick? What are the stimulants of progressiveness? . . . Some people suppose that the part of clock-winder is played by the inventor, the man of an original turn of mind, whose flow of new ideas and improvements keeps his firm constantly moving forward. This view is . . . out of date because, in the larger firms of today, the flow of new ideas for product and process innovation results, not from the chance inspiration of exceptional individuals, but from a deliberate decision by management to spend money on research and development.

Thus innovation was to be seen as demand-led. Demand-pull rather than technology-push interpretations of technological change dominated a number of other studies conducted during the 1960s. Such studies have been reviewed by, for example, Utterback (1974, 621), who writes:

> market factors appear to be the primary influence on innovation. From 60 to 80 per cent of important innovations in a large number of fields have been in response to market demands and needs. . .

while Gilpin (1975, 65) puts the point even more strongly:

> everything that we know about technological innovation points to the fact that user or market demand is the primary determinant of successful innovation. What is important is what consumers or producers need or want rather than the availability of technological options. Technological advance may be the necessary condition for technological innovation and on occasion new technology may create its own demand but in general and in the short-run, the sufficient condition for success is the structure or nature of demand.

The implication for R & D should then be a simple one: the pursuit of R & D is a consequence of a corporate requirement for innovation; innovation is not a consequence of the performance of R & D. However, matters may not be this simple. Mowery and Rosenberg (1979, 104) challenge this new orthodoxy, that 'the governing influence upon the innovation process is that of market demand; innovations are in some sense "called forth" or "triggered" in response to demands for the satisfaction of certain classes of "needs" '. They criticize the methodologies of a number of the studies that are frequently cited in support of this position. In particular, they argue (p. 140) that such studies have used the notion of market demand far too loosely: it should, strictly, be applied only to demand 'expressed in and mediated through the market place', and should be 'clearly distinguished from the potentially limitless set of human needs'. Secondly, Mowery and Rosenberg argue that interpreters of this body of research have failed to distinguish the differing roles for demand-side influence which have been attributed by different researchers. Langrish *et al.* (1972) refer to 'need-pull', and Myers and Marquis (1969) refer to 'demand' as a motivating influence. Project SAPPHO (Science Policy Research Unit, 1972) deals with the conditions underlying the commercial success of innovations, while Carter and Williams's study deals with the role of economic/market factors in the 'passage from invention to innovation'.

In their concern to re-establish the importance of supply-side considerations in the innovation process, Mowery and Rosenberg point to the rapid growth of the semiconductor electronics industry, pointing out (p. 148) that

the record of innovation in the industry has been one of sensitivity to the desires of customers (it is hard to see how matters could have been otherwise), particularly military customers. Yet, the case simply cannot be made that this demand for miniaturised solid-state components and logic functions was one which alone or primarily brought the industry into being.

We are reminded, then, that both technical and commercial potentials are required for successful innovation. This, of course, is not a new insight. Carter and Williams (1957) referred to the 'various necessary conditions of progress'. Langrish *et al.* (1972) explicitly rejected linear-sequential models and suggested (p. 57) that 'Perhaps the highest level generalisation that it is safe to make about technological innovation is that it must involve synthesis of some kind of need with some kind of technical possibility'. They further argued that alternative models need to be regarded as complementary rather than mutually exclusive, even when 'the complex process of innovation' is simplified by asking only: 'What stimulated the firm into the activity that led to the successful innovation?' This complementarity of demand and supply factors is discussed by Freeman (1982), who writes (p. 111) that the

> fascination of innovation lies in the fact that both the market and the technology are continually changing. Consequently, there is a kaleidoscopic succession of new possible combinations emerging. What is technically impossible today may be possible next year because of scientific advances in apparently unrelated fields . . . What cannot be sold now may be urgently needed by future generations.

Thus it might be supposed that the innovative firm will be constantly examining market possibilities in the light of (among many other things) available technology, and constantly examining technical possibilities in the light of (among many other things) available market opportunities. However, it may be preferable to refer to the demand side of technological innovation not just in terms of market opportunities but in terms of a wider range of economic demand factors, some indicated through the market mechanism and some not.

It is also ingenuous to suppose that the economic success of the individual firm in the industrial system demands a high level of innovativeness. Freeman (1982, 169) recognizes that 'rational profit maximising behaviour. . . is seldom possible in the face of the uncertainties associated with *individual* innovation projects', and goes on to suggest that several strategies are available to firms faced with the prospect of technical change. These range from 'offensive' innovation strategies 'designed to achieve technical and market leadership by being ahead of competitors in the introduction of new products' (p. 170), through 'defensive', 'imitative' and 'dependent' strategies to 'traditional' strategies, where the firm 'sees no reason to change its product because the market does not demand a change, and the competition does not compel it do so so' (p. 183), and to the 'opportunist' strategy based on 'imaginative entrepreneurship . . . which may bear little relation to R & D' (p. 183).

It is against this background of complexity that policy recommendations to increase the competitiveness and innovativeness of industry (whether at national or regional levels) must be judged. Clearly a simple exhortation to perform more R & D is as inappropriate now as it was when Williams (1967, 57) referred to the 'widespread belief that the speed of innovation is a simple function of research'. Yet innovation is plainly a function of research, if not a simple one; research is usually necessary for innovation if never sufficient, while the return to increased R & D activity will depend on a complex set of interactions and will thus vary from case to case.

Freeman associates his several innovation strategies with different types of business organization. This account has been drawn on by Taylor and Thrift (1983) in their elaboration of the segmentation of business organizations. They offer an account of the current pattern of segmentation which stresses, among other things, the relationship between segments, in particular that between smaller firms and larger business organizations. Such relationships may involve subcontracting, which is widespread in manufacturing industry, or franchising, which is more common in retailing but can also be found in the manufacturing sector.

It is clear that Freeman's 'ideal types' of innovation strategy can be associated with Taylor and Thrift's 'ideal types' of segmentation in a way which goes further than Taylor and Thrift have attempted. Indeed, Taylor and Thrift have used Freeman's categories to aid their general classification of the segmented economy. What is needed now, in order to aid the analysis of the segmentation (and eventually the location) of R & D, is the reverse association: an identification of certain innovation (and thus research) strategies with the 'ideal types' of the segmented economy.

A thorough attempt at this categorization is not attempted here. Rather, we will discuss just one such type—the 'laggard' company within the multidivisional corporate segment. Such a company might be classified in Freeman's typology as 'imitative' or 'dependent', with an emphasis on design and production rather than on research and development. It would be manufacturing products at a mature stage of the product cycle and its innovations would be mainly process ones. Such R & D as was performed would be concerned with the evaluation and implementation of these process innovations, but the firm would in most cases be a late adopter in the innovation process, and many of these evaluation and implementation functions would be adequately served by the sales and technical services organizations of its process plant suppliers. If levels of R & D were to be changed without otherwise changing the firm, the outcome in innovation terms would probably be negligible.

Plainly, this sort of argument needs further elaboration. In particular it is necessary to review more fully the literature on aspatial aspects of corporate innovation in order to identify and describe the association between 'technological role' and industrial segment. It is further necessary to identify the relationship between industrial segment and geographical space—a task which Taylor and Thrift identify in their suggested programme of research. We might tentatively suggest that spatial peripherality is linked through industrial segmentation to a technological role supporting an imitative or dependent innovation strategy. But the point on which we wish to insist for our present purpose is the simple one that low levels of R & D may in fact be the *optimum* levels for particular firms, given their technological roles in the segmented economy. Low levels of R & D are as likely to be the effect as they are to be the cause of low levels of innovative behaviour, and an apparent regional deficit of R & D might be explained by the fact that the *sorts* of firms which occupy peripheral locations simply do not *need* high levels of R & D. It is clear, furthermore, that exhortations to increase levels of corporate R & D may be misplaced. It is necessary to demonstrate that existing levels are indeed suboptimal and not just the consequence of the firm's adopted strategy. If low levels of R & D in peripheral regions reflect a regional specialization in corporate role within the industrial system, this may be unwelcome and efforts might be made to change it. But R & D levels will be the symptom and not the disease, and prescription should proceed accordingly.

5. Conclusion

Despite some early pioneering efforts, industrial R & D activity has not proved particularly amenable to geographical analysis and it has become increasingly apparent that a more thoroughgoing attempt by geographers to understand the *organizational processes* that are associated with R & D, together with an appreciation of R & D as part of the contemporary *political economy,* would both pay dividends. What has been attempted here has been a new interpretation of the geography of industrial research activity on this basis.

The higher profile accorded to the geography of R & D in the United Kingdom in the last ten years or so probably can be best explained by the recognition given to the role of industrial research in the process of restructuring during a period of change in the long waves of the economy (Hall, 1981). For governments intent upon understanding and directing such change this has meant supporting academic research in R & D and the wide field of 'technological change'. Whilst emphasizing the significance and importance of new industries and technologies, it would appear that some of this state support for such enquiries has been directed towards the fortunes of 'old' industries, often located in 'old' areas; for example, ACARD examined the coalmining machinery, rail transport, road construction, water supply and gas supply and distribution industries (Cabinet Office, 1980).

Another thrust of some of this research has been directed towards the less prosperous regions, working on the unproved assumption that if more R & D could be engendered therein or relocated to such areas from else-where, then perhaps new growth would be encouraged. Although some effort has been directed towards the spatial aspects of the R & D activity of the newer industries and their location (Thwaites, 1982; Oakey, 1984), similar examination of defence-dependent industries and government's own R & D appears to have been studiously avoided. The warnings to government about the United Kingdom's increasing technological con-centration on defence and nuclear industries (Pavitt, 1980) do not seem yet to have been heeded. Industrial economists have already pointed to the strategic implications and significance of this for the country's manu-facturing industries. To this warning the geographer and the regional analyst would add their advice that a continued concentration on in-dustries in the defence realm could lead to the further stagnation of the manufacturing economies of Britain's development areas.

What this chapter has tried to highlight has been the way in which defence-orientation has been paralleled by the effects of restructuring on British manufacturing industry—and especially on its R & D component —at a time of recession. The observations we make relate primarily to large corporations and to institutionalized R & D activity; whether our comments can apply to the small- and medium-size firm sector is open to debate. However, this concentration on the large corporation appears to be appropriate for two reasons: firstly, R & D and its geography does not appear to be behaving too differently from large-scale manufacturing as a whole; it is a product of a lengthy and continuing process of eco-nomic concentration. Secondly, a large proportion of the total R & D effort remains concentrated in the large firm sector. Although in the longer run the dynamic companies in the small-firm sector may grow, and in so doing help shape a new industrial geography, so far many would appear dependent upon their larger brethren. In order to draw attention to the dangers of oversimplifying sectoral or spatial relatonships, we have examined the work of Freeman (1982) and Taylor and Thrift (1983) in order to demonstrate the weaknesses of a centre-periphery model for the geography of R & D that does not include considerations of *segmentation*; the corporate organization's view of geographical space is not necessarily that of the geographer.

We have also attempted to stimulate interest in the role of *consumption* in the restructuring process as far as R & D is concerned. The relationship between demand for new and improved products and the research effort is still not fully understood, but it is our belief that access to newly evolving markets stimulates R & D (especially development work aimed at matching production to final demand) and thereby helps fashion its location. However, our conclusions here are more tentative because it would appear that large corporations can also be powerful agents in modifying markets to suit their products. Finally, we have tried to direct some attention towards the role of R & D in the service sector and suggest that industrial research is being used to alter the relations between labour and capital, usually to the detriment of the former in so far as the net balance of jobs is concerned. As the tertiary economy, and more especially the producer services of the quaternary sector, already shows evidence of spatial concentration in the core region(s) of states, we speculate on the extent to which the regional imbalance in the location of R & D activity is both the product of this process (especially in relation to headquarters' distributions) and the means whereby the patterns might or might not be continued.

Perhaps a new round of policy initiations needs to be encouraged not so much to stimulate R & D in those industries—often peripherally located—that currently appear to do little research because that might be by design, but rather towards shifting the balances in the research effort, especially from the military to the civil sector. At the same time, a keener appreciation by policy-makers of the role of R & D in corporate structures would reveal the futility of simply trying to encourage *more* research without attempting to restructure that industry in the light of regional capabilities and corporate objectives. These recommendations stem from our belief that the new structural interpretation for the location of R & D that we have outlined is one of the keys to the understanding of the relationships between research and development and regional development.

References

Blackburn, P., Green, K., and Liff, S. (1982). 'Science and technology in restructuring', *Capital and Class*, **18**, 15–37.

Brocard, M. (1981). 'Aménagement du territoire et développement régional: le cas de la recherche scientifique', *L'Espace géographique*, **6**, 61–73.

Buswell, R. J. (1972). 'Geography and the science of science', *Technology and Society*, **7**, 13–17.

Buswell, R. J. (1983). 'Research and development and regional development: a review', in A. Gillespie (ed.), *Technological Change and Regional Development* (London, Pion), pp. 9–22.

Buswell, R. J. and Lewis, E. W. (1970). 'The geographical distribution of industrial research activity in the United Kingdom', *Regional Studies*, **4**, 297–306.

Cabinet Office (1980). *Research and Development and Public Purchasing* (London, Cabinet Office).

Carter, C. F. and Williams, B. R. (1957). *Industry and Technical Progress: Factors Governing the Speed of Application of Science* (London, Oxford University Press).

Defence Committee (1982). 'Ministry of Defence organisation and procurement', *House of Commons Papers 1981-2* 22, I–II (London, HMSO).

Department of Trade and Industry (1983). *Regional Industrial Policy: Some Economic Issues* (London, Department of Trade and Industry).

Fothergill, S. and Gudgin, G. (1982). *Unequal Growth* (London, Heinemann).

Freeman, C. (1982). *The Economics of Industrial Innovation* (London, Frances Pinter), 2nd Edn.

Gershuny, J. (1978). *After Industrial Society: the Emerging Self-Service Economy* (London, Macmillan).

Gilpin, R. (1975). *Technology, Economic Growth and International Competitiveness. Study prepared for the Sub-committee on Economic Growth of the Congressional Joint Economic Committee* (Washington DC, US Government Printing Office).

Green, K. and Morphet, C. (1977). *Research and Technology as Economic Activities* (London, Butterworths).

Gudgin, G., Crum, R. and Bailey, S. (1979). 'White-collar employment in U.K. manufacturing industry', in P. W. Daniels (ed.) *Spatial Patterns of Office Growth and Location* (London, John Wiley), pp. 127–57.

Hall, P. (1981). 'The geography of the fifth Kondratieff cycle', *New Society*, 26 March, 535–7.

Holland, S. (1976). *Capital Versus the Regions* (London, Macmillan).

Howells, J. R. L. (1984). 'The location of research and development: some observatons and evidence from Britain', *Regional Studies*, **18**, 13–29.

Jordans, J. (1981). *The British Defence Industry* (London, Jordans and Sons (Surveys) Ltd.)

Keeble, D. (1976). *Industrial Location and Planning in the United Kingdom* (London, Methuen).

Keeble, D. (1980). 'Industrial decline, regional policy and the urban-rural manufacturing shift in the United Kingdom', *Environment and Planning A*, **12**, 945–62.

Kondratiev, N. (1935). Reprinted as 'The long-waves in economic life', *Lloyds Bank Review*, **129**, 41–60.

Langrish, J., Gibbons, M., Evans, W. G. and Jevons, F. R. (1972). *Wealth from Knowledge* (London, Macmillan).

Malecki, E. J. (1979). 'Locational trends in R & D by large U.S. corporations, 1965–1977', *Economic Geography*, **55**, 309–23.

Malecki, E. J. (1980). 'Corporate organization of R & D and the location of technological activities', *Regional Studies*, **14**, 219–39.

Malecki, E. J. (1983). 'Technology and regional development', *International Regional Science Review*, **8**, 89–125.

Mandel, E. (1975). *Late Capitalism* (London, New Left Books).

Massey, D. B. (1978). 'Capital and locational change: the U.K. electrical engineering and electronics industries', *Review of Radical Political Economy*, **10**, 39–54.

Massey, D. B. (1979). 'In what sense a regional problem?', *Regional Studies*, **13**, 233–43.

Massey, D. B. (1983). 'Industrial restructuring as class restructuring: production decentralization and local uniqueness', *Regional Studies*, **17**, 73–90.

Massey, D. B. and Meegan, R. A. (1979). 'The geography of industrial reorganisation: the spatial effects of the restructuring of the electrical engineering sector under the Industrial Reorganisation Corporation', *Progress in Planning*, **10**, 155–237.

Massey, D. B. and Meegan, R. A. (1982). *The Anatomy of Job Loss* (London, Methuen).

Ministry of Defence (1980). *Report of the Steering Group on Research and Development Establishments, under the Chairmanship of Lord Strathcona, Minister of State for Defence* (London, Ministry of Defence).

Mowery, D. and Rosenberg, N. (1979). 'The influence of market demand upon innovation: a critical review of some recent empirical studies', *Research Policy*, **8**, 102–53.

Myers, S. and Marquis, D. G. (1969). *Successful Industrial Innovation* (Washington DC, National Science Foundation).

NEDO (1983). *Civil Exploitation of Defence Technology* (London, NEDO).

Oakey, R. (1984). 'Innovation and regional growth in small high technology firms: evidence from Britain and the USA', *Regional Studies*, **18**, 237–52.

Pavitt, K. (1980). 'Industrial R & D and the British economic problem', *Research and Development Management*, **10**, 149–58.

Rothwell, R. (1982). 'The role of technology in industrial change: implications for regional policy', *Regional Studies*, **16**, 361–9.

Rothwell, R. and Zegveld, W. (1979). *Technical Change and Employment* (London, Frances Pinter).

Science Policy Research Unit (1972). *Success and Failure in Industrial Innovation* (London, Centre for the Study of Industrial Innovation).

Taylor, M. and Thrift, N. (1983). 'Business organization, segmentation and location, *Regional Studies*', **17**, 445–65.

Thwaites, A. (1982). 'Some evidence of regional variations in the introduction and diffusion of industrial products and processes within British manufacturing industry', *Regional Studies*, **16**, 371–81.

Thwaites, A., Oakey, R. and Nash, P. (1981). *Industrial Innovation and Regional Development. Final Report to the Department of the Environment.* (University of Newcastle upon Tyne, Centre for Urban and Regional Development Studies).

Townsend, A. R. and Peck, F. W. (1985). 'An approach to the analysis of redundancies in the U.K. (post 1976): some methodological problems and policy implications', in D. B. Massey and R. A. Meegan (eds), *The Politics of Method: Contrasting Studies in Industrial Geography* (London, Methuen).

Utterback, J. M. (1974). 'Innovation in industry and the diffusion of technology', *Science*, **183**, 620–6.

Watts, H. D. (1980). *The Large Industrial Enterprise* (London, Croom Helm).

Williams, B. R. (1967). *Technology, Investment and Growth* (London, Chapman & Hall).

Williams, T. I. (ed.) (1983). *Industrial Research in the United Kingdom* (Harlow, Longman), 10th Edn.

4 Research and Development in the Canadian Economy: Sectoral, Ownership, Locational, and Policy Issues*

JOHN N. H. BRITTON
Department of Geography, University of Toronto

1. Introduction

There have been notable increases in expenditures on Research and Development (R & D) in most advanced countries during the last two decades, but when expenditures are related to GDP, Canada is distinguished by the continuity of its position at the bottom of the OECD league table (Table 4.1). Through the 1960s and 1970s Canada's R & D effort hovered around the 1 per cent mark in contrast to the higher but often increasing allocations to R & D typical of its industrial competitors. Canada's low level of R & D activity has been explained by the lack of defence-related R & D, her specialization on resource production (primary processing activities are not research intensive), and her poorly structured industrial

Table 4.1 Canada's R & D* effort relative to selected OECD countries: selected years

	R & D/GDP (per cent)		
	1963	1971	1979
Canada	1.0	1.2	1.1
France	1.6	1.9	1.8
Germany	1.4[†]	2.1	2.4
Japan	1.3	1.6	2.0
Sweden	1.3[†]	1.5	1.9[§]
United Kingdom	2.3[†]	2.1[‡]	2.2
United States	2.9	2.6	2.4

* OECD statistics on R & D include Natural Sciences and Engineering (NSE) and Social Sciences and Humanities (SSH).
  † 1964.
  ‡ 1972.
  § Excludes SSH.
  Sources: OECD, *Technical Change and Economic Policy*, 1980 and *International Statistical Year 1979: Main Results*, March 1982.

* Editorial comments on an earlier draft by Jim Simmons and financial assistance from the Humanities and Social Sciences Committee (General Research Grant), University of Toronto are gratefully acknowledged.

economy. Nevertheless, the chief problem has been the federal government's perception of long-term policy needs. Canada's resource strength in international trade and employment growth in the late 1950s and 1960s masked for a long time the need to place a high government priority on industrial innovation. The motivation to address the technological weakness that derived from the high level of foreign investment in Canadian manufacturing was first realized in the 1970s, when the Foreign Investment Review Agency was established. In spite of that departure from the norm, until the current recession, Canadian socio-economic policies have concentrated on social issues and on redressing regional differences in economic growth, employment and income that have considerably more political significance within the confederation. Federal domestic policy has thus embodied substantial transfer payments to 'have not' provinces and even the incentives for industrial development (including new technology) have been a response to regional disparities with the further goal of expanding Canada's secondary manufacturing industrial base.

In 1978, after at least a decade of public reports decrying the paucity of Canada's R & D performance, the federal government announced a policy target for 1983 of 1.5 per cent of GNP. Although such a goal is not easily achieved, and the target date was later moved to 1985, the announcement was a victory for Canadians who have been concerned with the deplorable innovation performance of the Canadian economy. Since 1976 real R & D expenditures have increased by 55 per cent, despite the international recession that has reduced Canadian industrial development, and the proportion of industrial R & D has increased from 42 to 56 per cent.[1] Oil price increases, in particular, have stimulated R & D expenditures in the energy sector and this is a major source of the increase.

Despite the recent improvements in Canadian R & D activity, Canada's overall technological performance is low for a high wage, developed economy trading extensively in an international environment in which increasing value is attached to the technological content of goods and services. National trade indicators reveal that Canada is a net importer of finished manufactured goods, and relative to GNP this deficit increased over the 1970s as did the trade deficit on invisibles such as business services and technological inputs. By comparison internationally, Canada's trade performance deteriorated during the last decade in every category from primary products and food to manufactured goods, in which its share of world exports declined from 4.0 to 2.8 per cent (Science Council of Canada, 1981, 37). Clearly the trading value of Canada's resource staples is under pressure at a time when its technological performance cannot compensate. The technological trade deficit is a current reflection of Canada's limited past R & D effort, thus illustrating the long-term influence of R & D on national economic performance.

In order to establish the current pattern of Canada's R & D effort it is useful to distinguish between the funding and the performance of R & D by broad sectors—business, government, and universities—and to employ relative measures of R & D activity. The sectoral distinctions reveal the origins of Canada's weak international position: compared with other countries (Table 4.2) Canadian business participates in R & D to only a minor degree. Government *performance* of R & D in Canada lies in the middle of the OECD range but in terms of *government* funding, Canada is in last place: Canada's university sector is also in last place.

Despite absolute increases in R & D expenditures, Canada experienced decreases in performance of R & D in the 1970s (relative to the GDP) and the declines occurred in each of the three sectors—business 0.49 per cent GDP to 0.41 per cent, government 0.39 per cent to 0.27 per cent and universities 0.32 per cent to 0.24 per cent. This background suggests that a national target for R & D expenditures was a necessary step: the salient facts are that (1) despite Canada's modest expenditure on R & D in government laboratories, governmental support for R & D undertaken elsewhere in the economy, for example in the universities, is low by international standards; (2) both the funding and performance of R & D by business achieve an even worse international rating.

The depth of Canada's business funding problem may be illustrated by using the performance standards achieved in Sweden in 1979 (Table 4.2). Sweden is also a northern nation with significant resource sectors. If Sweden's 1.133 per cent GDP funding by industry were applied to Canada, the Canadian total R & D performance would increase to 1.725 per cent —the level of Japan, Sweden and France. This fanciful shift in business funding, however, would have required an 86 per cent increase of Canada's total R & D spending in 1979! While the business sector is Canada's major R & D problem it is also evident that government in Canada generates a much smaller direct flow of funds for business R & D than is true for all OECD countries other than Japan.

The scale of Canada's 'catch-up' problem is illustrated by Figure 4.1 which represents the announced targets for 1985 and the substantial changes in Canadian R & D performance that will be required to meet them. As proposed, the business share of R & D expenditure must increase in 1981-5 by 17 per cent per year in real terms, increasing its share of R & D spending from 36 to 50 per cent! The targets pose other serious problems:

Table 4.2 Canada's R & D by sector of performance/funding per cent GDP

	1971	1973	1975	1977	1979
Canada					
Bus.	0.489/0.383	0.399/0.310	0.416/0.341	0.397/0.321	0.413/0.334
Govt.	0.393/0.726	0.352/0.623	0.316/0.566	0.291/0.537	0.272/0.513
Univ.	0.324/0.056	0.258/0.041	0.259/0.049	0.252/0.047	0.241/0.045
France					
Bus.	1.070/0.700	1.036/0.681	1.075/0.705	1.061/0.723	1.076/0.780
Govt.	0.512/0.970	0.440/0.864	0.416/0.819	0.402/0.765	0.427/0.764
Univ.	0.297/0.157	0.277/0.161	0.288/0.177	0.273/0.163	0.280/0.159
United Kingdom					
Bus.	—	—	1.281/0.839	—	1.413/0.945
Govt.	—	—	0.551/1.124	—	0.468/1.059
Univ.	—	—	0.221/0.021	—	0.251/0.024

United States					
Bus.	—	1.616/1.012	1.572/1.028	1.573/1.035	1.597/1.089
Govt.	—	0.383/1.370	0.368/1.305	0.356/1.268	0.388/1.227
Univ.	—	0.354/0.017	0.361/0.016	0.347/0.015	0.344/0.012
Sweden					
Bus.	1.003/0.820	1.064/0.861	1.174/0.977	1.313/1.097	1.308/1.133
Govt.	0.127/0.600	0.132/0.670	0.137/0.669	0.158/0.707	0.160/0.712
Univ.	0.338/0.001	0.391/0.001	0.401/0.003	0.379/0.004	0.408/0.003
Japan					
Bus.	1.110/1.111	1.158/1.173	1.138/1.159	1.144/1.159	1.219/1.234
Govt.	0.236/0.549	0.274/0.557	0.246/0.596	0.239/0.585	0.259/0.627
Univ.	0.525/0.228	0.511/0.229	0.567/0.243	0.549/0.225	0.586/0.235
Germany					
Bus.	1.394/1.138	1.280/1.019	1.399/1.113	1.392/1.132	1.486/1.136
Govt.	0.311/1.018	0.341/1.041	0.367/1.054	0.344/0.947	0.389/1.070
Univ.	0.472/—	0.462/—	0.444/—	0.399/—	0.408/—

Source: Science Council of Canada, data from OECD.

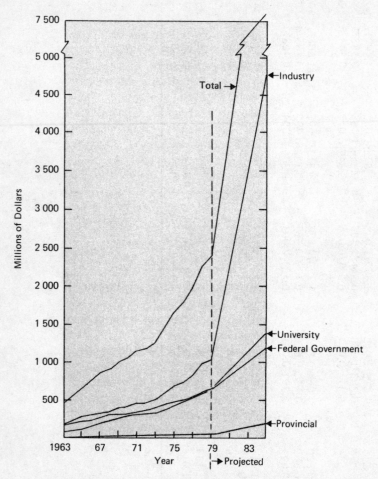

Figure 4.1 Actual and projected performance of R & D in Canada, 1963–85 *Source*: Science Council of Canada, 1981, 44 (data from Statistics Canada, Science Statistics Centre, and Ministry of State for Science and Technology).

Universities are to have a real annual growth rate of zero, to 1985, despite their pivotal role in supplying scientists and other educated workers, and in initiating, or pursuing, a significant portion of research;

R & D spending will require further capital investment for the downstream phases of the innovation process, but raising the required amounts of industrial working capital is an unfamiliar task for Canada's financial institutions. '$30 Billion will be needed to exploit the output'

(Science Council of Canada, 1981, 44) from the increased research program which, from Figure 4.1, will itself commit sums of the order of $7.5 billion per year to R & D;

Advocates of increased R & D would have preferred targets that are part of a total technological and industrial strategy. The full economic benefit of increases in R & D is likely to occur when industrial assistance policies for small enterprises and regional economic policy measures are in place, and when it is clear that downstream advantages of R & D undertaken by foreign subsidiaries in Canada benefit Canada and not only offshore parent firms. These policy and infrastructural elements are in flux, and none has a record of achievement that suggest they are ready to implement and channel the outcomes of the R & D stimuli to which the federal government is committed.

It would appear that Canada is entering a period of expanded technological expenditure as part of a federal government attempt to reposition Canada as a trading nation, the goal being to match the country's educational attainment with industrial and trade performance. That there will be difficulties is clear from what has been said, but to understand the problems that Canada faces in trying to meet its technological targets requires closer analysis of the determinants of the existing pattern of R & D. This paper begins with a sectoral perspective in order to specify Canadian and regional R & D specializations and to establish the impact on innovation of the high level of foreign ownership within manufacturing. The influence of size of firm on R & D activity is significant in the assessment of the impact of foreign ownership, and the necessity for an industrial system model is illustrated with respect to the international flows of industrial components and technological business services which are often a substitute for R & D in Canada. Attention is directed to the location of R & D throughout the Canadian metropolitan system in an analysis of (1) the national dominance of R & D by Toronto and Montreal, and (2) the sectoral and regional specialization of R & D activities shown by the industrial pattern of scientific and technological workers. Finally, government policy to stimulate industrial R & D is examined. Theoretical arguments in favour of government intervention are reviewed and direct grant and tax-based programmes are evaluated.

R & D expenditures are part of a process of industrial innovation through which technological capability is enhanced. So far in the discussion the standard international definition of R & D expenditures has been followed. As the chapter progresses, however, other forms of data are used that afford a broader evaluation of resources allocated to innovative activity. R & D expenditures and funding measure only activity in the earliest of the phases of the product life cycle and there are other expenses for design, engineering, production start-up, and marketing which are all costs of the *innovation process*. These factors are inadequately represented by R & D statistics alone.

2. Structural patterns of industrial R & D

2.1 *Industrial structure*

While the bulk of Canada's industrial R & D occurs in the manufacturing sector, as it does in other OECD countries, Canada has higher proportions of R & D in both the mining and service sectors, reflecting the importance of the resource sector (electric power utilities being included in services in Table 4.3). The weakest two economies in Table 4.3 in terms of manufacturing exports are Canada and Australia; both have developed on the basis of resource staples and foreign direct investment. In both cases manufacturing investment has been dedicated primarily to the domestic market, though in the last three decades Canada has become highly interdependent industrially with the US economy and has become identified with some technological specializations that have international markets. Within manufacturing, the chemical industry is inflated by the Canadian petroleum industry and this, plus the greater than average importance of R & D in the metals sector, reflects the structure of the Canadian resource base. Canada is strong in the electrical goods industry, particularly communications equipment, but obvious weaknesses exist in machinery and in transportation. The latter might seem to be surprising given the importance of the motor vehicle industry in Canadian exports and employment, but the Canada–United States auto-trade agreement has effectively restricted Canadian R & D in the assembly portion of that industry.

The trends in Canadian industrial R & D are easier to appraise if original Canadian data rather than the reprocessed OECD data are examined. In Table 4.4 gas and oil is the clear R & D leader but electrical products (especially communications equipment) and transportation equipment (dominantly aircraft and parts and transit systems) are also major sectors. Over the five-year period to 1980–2 (three-year average), industrial R & D in Canada increased by 60 per cent and differential growth was most marked in oil and gas, but electrical products and transportation equipment were not far behind.

Because the head office and laboratories of the oil and gas industry are spatially concentrated in Alberta—in spite of the exploration activity

Table 4.3 Industrial R & D in selected OECD countries, by industry: 1977 (per cent)

	Canada	United States	Sweden	France	Germany	Japan	Australia
Agriculture	n.a.	†	1.8	0.6	‡	0.2	n.a.
Mining	6.7	†	0.6	0.8	2.3	0.0	2.1
Manufacturing	77.8	96.8	89.6	93.0	92.2	91.2	63.3
Electrical	21.3	19.9	19.8	27.7*	26.2	23.8*	10.8
Chemical	16.9	14.0	11.6	18.3	26.9	19.6	13.8
Aircraft	11.2	23.7	§	18.6	7.2	§	§
Transport	1.7	11.6	20.3	11.9	12.2	17.0	7.5
Metals	8.5	3.1	8.3	3.7	2.8	8.2	11.8
Machinery	8.2*	18.0*	18.9*	5.0	14.1*	10.8	6.5*
Chemical-linked	4.6	3.2	4.1	5.6	2.0	5.6	8.6
Other	5.4	3.3	6.8	2.2	0.8	6.3	4.4
Services	15.5	3.2	7.8	5.6	3.5	8.2	34.6
Total	100.0	100.0	100.0	100.0	100.0	100.0	100.0

Note: Statistics are for 1977–8 in the cases of Canada, Sweden and Japan, and for 1976–7 in the case of Australia.
* Including computers.
† Included in services.
‡ Less than 0.005 per cent.
§ Aircraft included with transport.
n.a.—not available.
Source: OECD, *Science Resource Newsletter*, No. 5, Summer 1980 (from Treasurer of Ontario, 1983).

Table 4.4   Canadian industrial R & D expenditures, 1975-82

	1980-2* (1975 $m.)	Per cent	Increase 1975-7* to 1980-2 (1975 $m.)	Per cent
Mines and primary metal	82	7.3	11	2.6
Gas and oil wells, petroleum products	198	17.5	119	28.1
Food, beverage, tobacco	26	2.3	0	0
Wood-based industries	49	4.3	18	4.2
Machinery	79	7.0	25	5.9
Transport	168	14.9	84	19.8
Aircraft and parts	141		73	
Other transport	27		11	
Electrical	255	22.6	102	24.1
Communications equipment	216		90	
Other electrical	39		12	
Chemicals	88	7.8	20	4.7
Drugs	33		8	
Other chemicals	55		12	
Other manufacturing	48	4.2	11	2.6
Electric power utilities	62	5.5	21	5.0
Other non-manufacturing	75	6.6	12	2.8
Total	1,131	100.0	424	100.0

* Three year average.
Source: Statistics Canada, *Annual Review of Science Statistics 1982.*

in the Canadian north and eastern offshore areas—R & D expenditures are localized there as well. This is borne out by Table 4.5 showing that 66 per cent of R & D in the mines and wells sector occurs in Alberta; this is nearly half of Alberta's R & D and most of the balance comes from chemicals-based R & D. In this respect Alberta has been a location of growth. Nevertheless, Ontario performed 55 per cent and Quebec 24 per cent of Canada's total industrial R & D in 1981 and Alberta's share was only 13 per cent. The expenditure data, however, present only one measure of the level of R & D activity: the jobs generated by the pattern of expenditures introduces a major source of sectoral and regional variability that is determined by the organizational structure of R & D, sectoral cost patterns that are influenced by company size, as well as shifts in demand for personnel. For Canada as a whole, mines and wells generate nearly nine jobs per million dollars of R & D expenditure but, perhaps because of the speed of recent growth, Alberta's job ratio is less than seven jobs. In the electrical goods sector, by way of contrast, over nineteen jobs are created per million dollars of R & D expenditure in Canada and Ontario, with the largest segment of this industry, producing a job ratio of 18.5. The effect of differential employment ratios is to increase the 1981 importance of Ontario in Canadian R & D employment (62 per cent of person years) and to reduce that of Alberta (to 5 per cent) and Quebec (to 25 per cent). The importance of Canada's major manufacturing region in generating R & D is thus identified with the implication that corporate location and infrastructural factors, through the pattern of Canadian urbanization, are significant influences.

2.2 Industrial R & D and firm size structure

The dominant share of Canada's industrial R & D in 1981 was performed by firms with sales in excess of $50 million—82 per cent of expenditures being attributable to 28 per cent of firms. As noted previously, the business sector funds a substantial proportion of its own R & D expenditures and the incidence of company funding of R & D is directly related to company size, with only the largest-size class of companies exceeding the average level. These large firms generate 87 per cent of the total of company-funded R & D and, despite this, also absorb two-thirds of the funds that are derived from government. The data indicate that large firms are responsible for most of the R & D expenditures, but small firms contribute a higher proportion of their sales. The participant small firms, however, are surprisingly small in number and represent a tiny proportion of the large number of firms whose sales were less than $10

Table 4.5 Current intramural R & D expenditures by industry group and by region, Canada, 1981

Industry Group	Quebec		Ontario		Alberta		British Columbia		Other provinces*		Total	
	$m.	per cent	$m.	per cent	$m.	per cent	$m.	per cent	$m.	per cent	$m	per cent
Mines and wells	6	1.5	20	2.1	106	48.4	x	–	x	–	160	9.2
Chemical-based	61	14.9	190	20.1	93	42.5	3	4.1	31	36.5	377	21.8
Wood-based	27	6.6	19	2.0	–	–	18	24.7	2	2.4	67	3.9
Metals	24	5.9	70	7.4	x	–	1	1.4	x	–	100	5.8
Machinery and transportation equipment	168	41.1	199	21.1	1	0.5	4	5.5	17	20.0	389	22.5
Electrical	73	17.8	304	32.2	4	1.8	13	17.8	4	4.7	398	23.0
Other manufacturing	3	0.7	15	1.6	x	–	x	–	–	–	19	1.1
Services	48	11.7	128	13.5	14	6.4	20	27.4	11	12.9	221	12.8
Total	409	100.0	945	100.0	219	100.0	73	100.0	85	100.0	1730	100.0

* Includes the Yukon and the Northwest Territories.
x—confidential.
Source: Statistics Canada, *Standard Industrial R & D Tables 1963–1983.*

million in 1981 (Table 4.6). It should be noted that manufacturing firms with sales up to $20 million (1978) comprised 98 per cent of the population of manufacturers although they generated only 26 per cent of sales (see Table 4.7).

The small number of innovation-orientated manufacturing firms in Canada is a problem that has been confronted by a number of scholars. Steed (1982), for example, was able to identify only 165 'threshold' firms of small or medium size and these fell into the largest two size classes of Table 4.6. The smallest plant Steed considered has an employment of 100 because he reasoned that the scale of an enterprise is a determinant of its ability to finance R & D and in this he is supported by other scholars: Crookell *et al.*, for example, have concluded that the limited size of the Canadian market inhibits the ability of small firms to generate new products on a continuous basis (Steed, 1982, 38-9). Steed concludes that a strikingly small number of threshold firms have achieved sufficient profitability to constitute a vital and innovative core (Steed, 1982, 137). He echoes a notion developed in Britton and Gilmour (1978) that to expand the market for small indigenous firms will require the development of more, larger Canadian (core) companies. The reasoning behind this argument involves the impact that backward linkages from larger manufacturing firms can have on small firms induced to respond technologically to the product innovations of the larger firms. While positive instances exist in Canadian manufacturing, for example, in the auto parts industry where survival has required participation in 'down-sizing' and weight reduction, generally it has been found that the linkage structure of the Canadian industrial system contains serious faults (Britton and Gilmour, 1978; Britton, 1974, 1976). The potential for innovative R & D-generating firms is thus significantly reduced.

3. Foreign ownership and the Canadian industrial system

3.1 *Foreign firms and Canadian R & D*

The low level of Canadian industrial R & D and the poor performance of Canada in the technology-intensive sphere of international trade (an import penetration of 31 per cent) result from the inadequate R & D expenditures of foreign-owned firms and their lack of participation in Canadian innovation, exports and backward linkages to Canadian enterprises. Canada, it should be noted, is distinctive in its high level of foreign direct investment. In the highly technology-intensive sector, foreign-owned firms account for about 70 per cent of the sales of manufacturing firms, and their dominant position in these industries is illustrated in Table 4.8.

Table 4.6 Canada: industrial R & D by company size, 1981

Company sales ($m.)	No firms*	Total R & D funds spent ($ m.)	Company R & D funds spent ($ m.)	Per cent R & D company funded	Company R & D funding/$100 sales
Less than 1	183	39	13	33.3	5.0
1–9.9	332	110	58	52.7	1.9
10–49.9	281	205	128	62.4	0.8
more than 50	306	1,606	1,301	81.0	0.9
Total	1,102	1,960	1,500	75.0	0.9

* Excludes non-commercial companies.
Source: Statistics Canada, op. cit. reporting results of the 1981 Industrial R & D Survey: 5,286 respondents (73 per cent rate of success), only data on firms who reported undertaking R & D are published.

Table 4.7   Size distribution of Canadian manufacturing businesses; sales classes, 1978

Sales, $ thousands	Businesses		Total sales		Average no. employees
	No.	Per cent	$ million	Per cent	
Less than $50	13,450	29.6	289.9	0.2	2
$50–$249	13,480	29.7	1,659.4	1.1	6
$250–$1,999	12,556	27.7	9,658.6	6.2	23
$2,000–$20,000	4,981	11.0	28,757.7	18.6	123
More than $20,000	921	2.0	114,275.1	73.9	1,511
Total	45,388	100.0	154,640.8	100.0	53

Source: Minister of State, Small Business and Tourism, *Small Business in Canada*, 1981.

Table 4.8   Canadian and foreign-controlled proportions of sales and R & D: research-intensive industries, 1975

Technology Intensity	Canadian-controlled		Foreign-controlled	
	Sales (%)	R & D (%)	Sales (%)	R & D (%)
Medium Tech				
Pulp and paper	56.4	67.2	43.6	32.8
Primary metals	82.9	86.0	17.1	14.0
Ferrous	87.0	88.8	13.0	11.2
Non-ferrous	78.6	85.2	21.4	14.8
High Tech				
Electrical	34.4	59.2	65.6	40.8
Machinery	32.5	31.4	67.5	68.6
Business machines	14.8	11.3	85.2	88.7
Other machinery	35.0	43.8	65.0	56.2
Chemicals	17.1	31.7	82.9	68.3
Pharmaceuticals	13.2	29.3	86.8	70.7
Other chemicals	18.3	33.2	81.7	66.8
Aircraft parts	17.3	41.9	82.7	58.1

Source: Ministry of State for Science and Technology, *R & D in Canadian and Foreign-Controlled Manufacturing Firms,* Background Paper No. 9, 1979.

Through their behaviour, foreign-owned firms have a disproportionate influence on all trade and industry performance measures. While most foreign-owned firms provide limited propulsive influence through backward linkages to smaller, Canadian firms, in the services or manufacturing (Britton, 1974), domestic enterprises themselves have not been as aggressive and innovative as opportunities allowed. In the 1960s, for example, Canadian firms experienced major expansion of the domestic market and foreign opportunities which could have stimulated firms to invest in R & D, technological upgrading, product innovation and market exploration. These opportunities, and the international example of firms of other countries, have had modest impact on firms of domestic origin. Nevertheless, the absence of both a strong entrepreneurial tradition in manufacturing (in part because of fragmented markets), and a supportive capital market, have been recognized in Canadian research as additional significant impediments to domestically-owned industrialization. As a consequence only a small number of innovative Canadian firms were identified by Steed as 'threshold'. Furthermore, the fact that only a few Canadian firms themselves attain both the scale and the level of innovation to act in a propulsive manner to the benefit of small enterprises is itself an impediment to increased R & D and further industrial innovation. As industrial economies develop technologically, intersectoral interdependence should increase as well. The manufacturing sector in Canada, however, generates a circumscribed demand for producer services and industrial components because of the high level of imports of services, goods for resale, and components that are maintained by foreign-owned firms. The evidence (Table 4.9) shows that the import propensity of foreign-owned manufacturers (for goods) is nearly four times that of domestic industry (2.5 times when auto industry imports are excluded). In all, 71 per cent of non-auto industry imports are made by foreign subsidiaries. The highest import propensities, however, are found in the technology-intensive industries that have a high level of foreign ownership. Canada also imports a substantial quantity of business services, mainly by foreign subsidiaries, and while the data are less detailed, business services add another 10 per cent in payments to the $29 billion of commodity imports.

These industrial imports contain some specialized inputs otherwise not available to the Canadian manufacturing sector: nevertheless, they also represent foregone opportunities for domestic producers to achieve thresholds for start-up and for existing enterprises to meet the demands

Table 4.9   Merchandise imports by manufacturing industries, 1978

| | Import propensity* | | | | Imports by foreign-controlled companies (%) |
	Foreign-owned (%)	Canadian-owned (%)	Total (%)	Foreign control (sales)	
Automobile industry	60.8	8.6	58.5	95.6	99.4
Non-auto manufacturing	18.5	7.8	13.2	50.1	70.5
Total manufacturing	28.9	7.8	19.8	56.7	82.9
Technology-intensive industries					
Machinery	39.5	17.8	32.0	65.5	80.8
Non-auto transport	28.6	17.4	22.6	46.8	59.1
Electrical products	24.4	9.9	19.9	69.0	84.6
Chemicals	22.1	5.2	18.0	75.3	92.8
Total	27.4	12.0	22.5	68.2	83.1

* Imports/$100 sales.
Source: Statistics Canada, 1978, 67–509.

Table 4.10 Canada: in-house R & D intensities:* Canadian and foreign-controlled firms, Canada, 1973 and 1979

	1973		1979	
	Canadian-controlled firms (%)	Foreign-controlled firms (%)	Canadian-controlled firms (%)	Foreign-controlled firms (%)
Mines	x	x	0.76	0.75
Gas and oil wells	x	x	1.23	0.84
All mining	0.87	0.63	1.05	0.81
Food, beverages and tobacco	0.21	0.39	0.16	0.40
Rubber and plastic products	1.31	0.65	1.08	0.73
Textiles	0.41	0.62	1.87	0.58
Wood-based industries	0.25	0.37	0.27	0.25
Primary metals (ferrous)	x	x	x	x
Primary metals (non-ferrous)	0.74	1.39	x	x
Metal fabricating	0.77	0.33	0.33	0.37
Business machines	16.92	1.63	7.99	1.33
Other machinery	1.61	1.10	1.08	0.65
Aircraft parts	5.40	18.05	19.03	5.22

Other transport equipment	2.54	0.13	2.11	0.22
Other electrical products	1.20	1.36	1.03	1.06
Communications equipment	7.96	4.51	9.73	5.53
Non-metallic mineral products	0.47	0.36	0.83	0.48
Petroleum products	x	x	0.02	0.39
Drugs and medicine	14.60	3.59	7.51	3.45
Other chemical products	1.65	0.99	2.59	0.68
Scientific professional equipment	19.93	0.84	17.72	0.66
Other manufacturing industries	0.47	0.70	0.75	0.22
All manufacturing	0.97	0.91	1.06	0.68
Transportation and other utilities	x	x	0.24	—
Electrical power	0.59	x	0.78	—
Engineering and scientific services	4.39	5.68	7.76	5.84
Other non-manufacturing industries	x	x	2.41	1.00
All services	0.55	0.53	0.61	2.29
All industries	0.88	0.89	0.92	0.69

* R & D expenditure/$100 sales.
x—confidential.
Source: Data provided by the Science Centre, Statistics Canada, Economic Council of Canada, 1983, 42.

of Canadian industry for high-value inputs. In essence, the openness of the Canadian industrial system to foreign direct investment generates a high level of imports of intermediate goods and services, which are often intracorporate flows, depressing the development of Canadian supply firms. The feedback effect on innovation in existing domestic enterprises and on foreign firms interested in obtaining Canadian inputs is obvious.

It is argued here that the degree of foreign control in the manufacturing sector is a strong and directly negative influence on Canada's R & D performance, especially in industries that rely on R & D for survival in international markets. The broad outline of the argument is supported by the leadership of domestic over foreign-controlled firms in R & D intensities (Table 4.10), intensities that increased considerably throughout the 1970s. In the research on this apparent difference in behaviour, Frankl's work (1979) has been of particular value in verifying a variety of relationships involving Canada's R & D and the impact of foreign ownership. Using data for 115 industries for 1972, she established that the degree of foreign control in an industry has a statistically significant negative effect on the R & D intensity of an industry in Canada relative to the R & D intensity of the industry in the USA (Frankl, 1979, 2). Foreign subsidiaries can easily import the results of R & D within the structure of the corporation when it is needed, and this is a likely explanation of the pattern. It is consistent that the Canadian-controlled segments of most industries attain a higher level of R & D intensity than the foreign-controlled counterparts. Limited R & D functions within subsidiaries also agree with a pattern of corporate organization in technology-intensive industries that reaps rewards through economies of scale from the centralization of strategic activities such as R & D. The importance of the lower R & D intensity recorded by foreign-controlled firms is made clearer when the R & D intensity for the USA and that for Canadian-controlled companies are used as hypothetical replacements. These substitutions go some way towards permitting a glimpse of the developmental cost of the high level of foreign penetration in technology-intensive industries. In fact the total R & D by foreign-controlled companies would be almost *tripled* if US levels were in force and would lead to a *doubling* of the total of R & D in Canada. Similarly, the R & D levels of Canadian-controlled firms would, if substituted, increase the R & D performance of foreign subsidiaries by about 130 per cent and increase total Canadian R & D by 65 per cent.

Frankl shows that foreign-controlled firms in the automotive assembly industry perform a very low level of R & D. If these firms performed at the R & D intensity of Canadian-controlled companies in the sector (parts producers), their R & D expenditures would have been 6,800 per cent larger (in 1972)—nearly 40 per cent of the total shortfall in industrial R & D. If levels of R & D intensity for the US parents prevailed in the foreign-controlled plants of the auto sector, nearly 50 per cent of the total imputed R & D gap would be eliminated!

The low levels of R & D intensity in the auto industry are attributable to the corporate arrangements that have been produced under the Canada–US Auto Pact through which only assembly functions have been assigned to Canadian plants of the Big Four Corporations. In the aircraft sector, however, the foreign-controlled firms have maintained a high level of R & D intensity while Canadian-controlled companies lag behind. The high degree of world product-mandating and the defence-sharing and defence-subsidizing programmes of the federal government are the probable explanations.

Frankl's research is an important element in understanding the R & D performance of Canadian industry. Nevertheless, none of her results produced statistical significance for the size of firms as an influence on R & D intensity. In addition, collinearity was identified between ownership and company size and the logical suggestion is made that plant-level data are required to evaluate company size as a determinant of R & D intensity. The hypothesized negative influence of foreign ownership on R & D intensity has been pursued by others, however, in the context of research into the effect of size of firm.

When viewed across the size distribution of firms undertaking R & D, the pattern of R & D intensity (R & D expenditures/sales) in Canada follows an inverse relationship which is at variance with the conventional observation of either increasing intensity or an inverted U-shaped relationship (Table 4.11). There is a high proportion of Canadian-owned firms in the 'small' size class—less than 30 per cent foreign-owned—and a high proportion of foreign-owned firms in the category of 'large' R & D performers—more than 60 per cent foreign-owned—(Chand, 1980, 23). When these two pieces of information are related back to the data of Table 4.8, the influence of the low level of R & D intensity of foreign-owned firms is apparent. The 1983 Ontario Budget Paper concludes that foreign firms draw on the technology of their parents allowing them to 'perform significantly less R & D relative to sales than their Canadian counterparts . . . the available evidence supports the view that foreign-owned firms allocate fewer resources to R & D than resident-owned firms of comparable size facing similar economic circumstances' (Treasurer of Ontario, 1983, 11–12). The evidence, while not abundant, is persuasive,

Table 4.11 Canada: company-funded R & D intensity:* manufacturing companies, 1975

Industry	Small[†]	Medium[‡]	Large[§]	Total
Food, beverages and tobacco	5.02	0.80	0.17	0.25
Rubber and plastic products	2.88	0.96	0.41	0.51
Textiles	1.41	2.48	x	0.70
Wood-based	3.09	0.89	0.29	0.33
Primary metal	x	1.15	0.65	0.66
Metal fabricating	3.17	0.60	0.24	0.47
Machinery	4.05	2.54	0.50	1.03
Transportation	4.89	1.94	0.36	0.49
Electrical products[‖]	4.28	2.50	2.32	2.48
Non-metallic mineral products	3.36	x	0.26	0.31
Chemical products	3.06	2.37	0.97	1.37
Other manufacturing	2.73	0.88	0.12	1.12
Total manufacturing	3.65	1.74	0.55	0.71

* R & D expenditures/$100 sales.
† Sales of less than $10 million
‡ Sales of $10–£50 million
§ Sales of over $50 million
‖ Includes scientific and professional equipment
  x––confidential.
Source: Chand, 1980, 21.

and corroboration of this conclusion is provided by de Melto *et al.* (1980) who present R & D data for employment size groups for Canadian and foreign-controlled firms. In a sample of 134 innovators in medium- and high-technology industries, both small and large Canadian-owned firms have higher R & D intensities than foreign-controlled firms (Table 4.12).

Although the distribution of foreign-owned companies is biased in favour of larger enterprises, and although company size and foreign ownership are collinear variables, several studies have proceeded as if there is no statistical problem and have concluded that foreign ownership is not a significant influence on R & D intensity. Regressions of R & D outlays on quadratic or cubic polynomial forms of company sales effectively mask the effect of ownership in telecommunications (Cohen *et al.,* 1984) and other technology-intensive industries (Hewitt, 1981). Thus, the use of data for firms has not avoided the modelling problems first encountered with aggregate data.

Table 4.12 Average R & D/sales in Canadian foreign-controlled firms, 1978

No. of employees in the field	Canadian-controlled firms		Foreign-controlled firms	
	No.	R & D/sales (Mean %)	No.	R & D/sales (Mean %)
50 or less	34	11.2	13	3.3
100 or less	45	10.1	23	4.3
200 or less	54	9.1	38	3.7
500 or less	60	8.4	54	3.4
More than 500	5	10.3	14	2.0

Source: de Melto *et al.*, 1980, 44.

3.2 *The political–economic debate*

The R & D performance of foreign-owned firms in Canada is a highly charged political–economic issue. It is not denied that foreign subsidiaries act as a pipeline for technology made available by their parents and that there are short-run gains in access to domestically produced goods from this trade. In the longer run, however, significant disadvantages derive from an overdependence on imports of corporate R & D; much R & D performed in Canada is being devoted to adapting foreign technology to suit the Canadian market, and there has been a tendency for subsidiaries to maintain only satellite or technical service labs, and to be unconcerned with exporting. 'Learning by doing' within the Canadian labour force cannot be guaranteed if technology is imported and employment opportunities are reduced if technology which could have been produced domestically is imported.

Long-term gains, however, are recognized when subsidiaries are allowed to develop their own R & D and product specializations with the goal of exporting. These North American or world product mandates are known to stimulate Canadian R & D and thus open up possibilities for long-term development. Governments are clearly working towards higher levels of export mandating among subsidiaries and it is hoped that the writers of the Ontario Budget Paper (1983) have data which support their claim that 'over the last decade, a decline in the relative importance of foreign ownership in Canadian industry has been accompanied by an increase in the extent to which foreign subsidiaries compete in global markets. As these changes take root throughout the economy, foreign-owned firms operating in Canada will make a greater contribution to industrial innovation than in the past' (Treasurer of Ontario, 1983, 12).

Much of the debate over the low level of R & D undertaken by foreign subsidiaries has focused on policy goals such as technological sovereignty (to combat the effects of foreign direct investment). One group of writers (Britton and Gilmour, 1978; Science Council of Canada, 1979; Hayter, 1982a) argues that there are significant developmental losses that stem from low levels of R & D by foreign subsidiaries because of the associated impacts of:

1. the small number of world product mandates that generate Canadian exports from Canadian innovation;
2. the substantial technological imports of specifications, licences, patents and professional services;
3. the technological imports that are embodied in the capital goods, sub-assemblies, components and consumer products that are imported by foreign subsidiaries; and
4. the small number of opportunities for technological spread effects that derive from foreign firms because of their limited backward linkages into the domestic economy.

These aspects of Canada's technological dependence were perhaps inevitable given the century-long policy environment of protection-via-tariffs. Nevertheless, the Science Council has proposed a number of policy measures to develop technological sovereignty by strengthening the innovative capacity of Canadian firms in the setting of falling tariffs created by the last round of GATT. The policy recommendations advanced by the Council have promoted strong opposition from some economists who believe in free international markets and less intervention by governments into the domestic operations of the economy. Their case assumes that science, technology and R & D policies will be inefficient, or at least inequitable, favouring one sector at the risk of global welfare maximization. In the final analysis, the critics believe in the operation of 'the market process' without the intervention of governments. Their case is faulted, however, because they do not recognize that, in a world of heavy defence spending, large government procurement contracts, environmental and business regulation, and outright grants and production and export subsidies, 'the international market' is a political process as much as an economic entity. The neo-classical position sees no place for national R & D goals; for example, the notion that current increases in R & D are necessary for longer-term development is dismissed as a naive growth model (Rugman, 1981, 607), neglecting the fact that most Western economies are following this particular path towards techno-economic development with no more sophistication and much more resolve than has been true for Canada. Canadian government support for R & D, when measured in dollar commitments, is small by international standards. Still, Rugman comments that 'it is, of course, not clear that there is anything suboptimal about the current ratio of R & D to sales' (ibid., 607). He believes that policy initiatives designed to generate technological competence in Canadian industry are inefficient, even though they are directed to ensuring that the economy enjoys technological *interdependence* (not independence and not dependence) with the rest of the world. In this respect Rugman and others such as Daly and Safarian are all blinded by a paradigm that seeks efficiency of the firm (or within the firm) yet which has nothing progressive to say about Canada's industrial performance, structure, or policy (Daly, 1979; Safarian, 1979). These scholars cling to neo-classical constructs but write on corporate behaviour and government programmes. Their mixture of theory and reality is not creative in the Canadian situation since it does not have roots in a political–economic conception of Canada's difficulties.

4. The localization of Canadian R & D

Canadian R & D is highly concentrated in the central and industrial provinces of Ontario and Quebec, although the growth of the petroleum industry has generated new R & D locations in Alberta. There are no data on the inter-urban distribution of R & D in Canada, but there is a variety of surrogate measures; for example, the twenty-four metropolitan centres of over 100,000 population (census metropolitan areas—CMAs) accounted for 73 per cent of the 1981 jobs in the occupations relevant to R & D—mainly natural scientists, engineers and mathematicians (NSEM). These urban centres include 56 per cent of Canada's population and 59 per cent of her labour force. While NSEM data confirm the importance of metropolitan locations of technological activity, they also indicate that technological occupations are locationally concentrated within the metropolitan system: two centres generate 30 per cent of the jobs and another five centres the next 50 per cent. Despite this, only two centres demonstrate a clear degree of localization[2] of NSEM jobs—Ottawa-Hull and Calgary. Given the location of federal government jobs in Ottawa, the localization of NSEM jobs in Ottawa-Hull is not surprising. In the case of Calgary, however, its employment structure experienced significant changes between 1971 and 1981 and it is necessary to describe the pattern of Canadian metropolitan change as

a context for the examination of the locational pattern and changes in
R & D-related jobs in Canada.

During the 1970s, Canada experienced two substantial shocks to its
urban system. First, the international repricing of oil generated significant
new exploration activity with an associated demand for consultants and
staff. The result was an expansion in jobs in Alberta and an over-response
in migration into Edmonton and Calgary. Second, the 'Séparatiste' movement in Quebec and its political–economic implications (as perceived by
Anglophones and international business) and the election of the Parti
Québecois government produced a pattern of selective emigration which
slowed the growth of Montreal. These significant events disturbed an
otherwise more predictable set of changes in the metropolitan system
which reflect a declining birth rate, a reduction in international immigra-
tion, and internal migration dampened by the recession. The major net
effects of all the impacts on metropolitan population 1971–81 are described in Table 4.13. Toronto gained on Montreal and experienced a
modest population increase of 14 per cent; Edmonton and Calgary,
however, experienced large increases in population (of 33 and 47 per cent
respectively). The *labour force* of metropolitan centres proves to be an
even more dramatic measure of economic change in the decade 1971–81
since the 'baby boom' cohort began to move into the work-force. Over
the period, the Canadian labour force increased by 42 per cent and this
was reflected across the set of metropolitan centres. Edmonton and
Calgary, however, advanced their ranking within the metropolitan system
(Table 4.14) with increases of 73 and 98 per cent respectively. Although

Table 4.13 Canada: metropolitan* population rank change, 1971–81

Popn. '000	1971 Rank	1981 Rank	Popn. '000
2,734	1. Montreal	1. Toronto	2,999
2,628	2. Toronto	2. Montreal	2,828
1,082	3. Vancouver	3. Vancouver	1,268
603	4. Ottawa-Hull	4. Ottawa-Hull	718
540	5. Winnipeg	5. Edmonton	657
499	6. Hamilton	6. Calgary	593
496	7. Edmonton	7. Winnipeg	585
481	8. Quebec City	8. Quebec City	576
403	9. Calgary	9. Hamilton	542
303	10. St. Catharine's	10. St. Catharine's	304

* CMAS with populations of 500,000 only.

Table 4.14 Canada: metropolitan labour force rank change, 1971–81

Labour force '000	1971 Rank	1981 Rank	Labour force '000
1,245	1. Toronto	1. Toronto	1,705
1,080	2. Montreal	2. Montreal	1,451
475	3. Vancouver	3. Vancouver	691
260	4. Ottawa-Hull	4. Ottawa-Hull	393
244	5. Winnipeg	5. Edmonton	378
219	6. Edmonton	6. Calgary	352
213	7. Hamilton	7. Winnipeg	314
178	8. Quebec City	8. Quebec City	285
178	9. Calgary	9. Hamilton	284
130	10. London	10. Kitchener	
125	11. St. Catharine's	Waterloo	156
	12. Kitchener	11. London	155
106	Waterloo	12. St. Catharine's	152
105	13. Windsor	13. Halifax	146
96	14. Halifax	14. Windsor	119
81	15. Victoria	15. Victoria	119
62	16. Sudbury	16. Regina	88
62	17. Regina	17. Saskatoon	83
53	18. Saskatoon	18. Oshawa	
48	19. St. John's	19. St. John's	73
47	20. Thunder Bay	20. Sudbury	71
43	21. Saint John	21. Thunder Bay	64
	22. Chicoutimi-	22. Chicoutimi-	
40	Jonquière	Jonquière	59
		23. Saint John	53

absolute increases of nearly 160,000 and 174,000 produced these changes, Toronto's absolute increase was substantial at 460,000, indexing the degree of stability in the system.

Within this context the growth of NSEM 1971–81 follows strongly the growth pattern of the whole labour force.[3] Calgary, however, has a large positive deviation (more than 2 standard errors) while smaller deviations occur for Montreal and Victoria (negative) and Vancouver (positive). By 1981 Calgary's NSEM labour force exceeded that of Ottawa-Hull. Toronto and Montreal maintained a dominant position, but these two centres failed to match their 1971 share of NSEM jobs in their 1971–81 growth. Nevertheless, the growth of the largest five centres, Toronto, Montreal, Vancouver, Ottawa-Hull, and Calgary (54.3 per cent of national growth), exceeded the 1971 share of this group of centres (53.5 per cent).

The international literature (Maleki, 1983) suggests that the location pattern of the NSEM labour force and its recent growth should reflect the influence of corporate head offices, research universities, and government scientific establishments. Among the major centres, Ottawa-Hull's labour force is influenced by the location of laboratories in federal government departments and auxiliaries and 53 per cent of the 'person years' in federal scientific establishments (57 per cent of their expenditures) are located in this one metropolitan centre. By way of contrast, head offices and university research activity (Table 4.15) all show a pattern of extreme spatial concentration in Toronto and Montreal, thus implying that the conventional determinants of R & D (measured in this case by the relevant scientific and technical jobs) operate in the Canadian economy. Nevertheless, the concentration of NSEM jobs and NSEM growth in Toronto is weak when compared with the measures of head office location in Toronto (Table 4.15).

In light of the size and manner of impact of foreign direct investment on the Canadian industrial system, the geographic pattern of NSEM jobs may reflect the metropolitan incidence of foreign control. Is there a reduced level of R & D activity in locations that have large proportions of their head office and/or manufacturing production in foreign-controlled companies? An explanation is also sought for the limited impact that resource-related R & D has on centres such as Vancouver despite their favourable location with respect to the resource economy.

In order to explore these concerns, recourse is made to data that exclude government laboratories, universities, and other non-profit institutions which inevitably influence the pattern of the NSEM labour force. The R & D units and the labour force engaged in scientific and technological (S & T) work in the private sector are given in a directory of business units pursuing this type of activity (MOSST, 1978). The directory is based on a 46 per cent response rate to a survey of potential respondents; for each business unit, in 1977, the labour force, industry and untied (open) or tied (inhouse) markets for its services are specified. Overall there is a strong locational relationship between the S & T workforce for 1977 and the NSEM data for 1981.[4] Negative deviations from the general relationship reflect the relatively modest private sector S & T functions in a number of provincial capitals but this does not apply to Toronto, Edmonton or to Ottawa, the federal capital.

Nearly half the S & T jobs[5] are in the producer service sector—predominantly engineering and scientific services (75 per cent) and computer services (18 per cent)—with only 4 per cent of these 'in-house' jobs (Table 4.16)! By contrast, no other sector has less than 27 per cent

Table 4.15 Canada: indicators of metropolitan functions: selected centres (per cent)

	Toronto	Montreal	Vancouver	Ottawa-Hull	Calgary	Edmonton
Head Office revenue, non-financial corporations (1977)*						
Domestic	39	32	7	2	5	1
Foreign	57	17	4	—	5	—
Total	48	26	6	1	5	1
Ph.D. enrolment in Natural sciences and engineering 1981	26	21	9	6	2	1
Expenditures of federal R & D establishments 1981–2	3	3	1	57	1	—
Share of NSEM in labour force 1981	18	13	6	6	7	4
Share of growth of 1971–81 NSEM in labour force	16	10	7	6	10	6

* Manufacturing, Service, and Resource Corporations: proportion of total revenue are 47 per cent, 35 per cent and 18 per cent respectively.

Source: Science Statistics Centre *Federal Scientific Establishments 1981–82*; Canadian Association of Graduate Schools, *1982 Statistical Report* (Council of Ontario Universities, 1982); Semple and Smith (1981), 4–26.

Table 4.16 Canada: survey of scientific and technological capabilities labour force by sector, 1977

	Tied per cent	Untied per cent	Sectoral labour force
Producer services	4	96	48
Technology-intensive manufacturing	40	60	17
Other secondary manufacturing	32	68	22
Resource and primary processing	27	73	13
Total labour force	(9,019) 19	(37,490) 81	(46,509) 100

Source: MOSST, 1978.

'in-house'. The S & T units in the producer service sector also are larger than those in the production sectors (30 vs. 18 employees): this means that despite its resource specialization Canada does not generate many large corporate R & D laboratories. Hayter, for example, shows that US establishments in the forest product sector have much larger R & D establishments than are found in Canada (Hayter, 1982b).

Toronto and Montreal generate 57 per cent of Canada's private sector S & T jobs, a proportion that better reflects the importance of their head office functions than shown in the more broadly defined NSEM data (Table 4.17). Technology-intensive manufacturing is most highly concentrated in Toronto and Montreal while the resource and primary manufacturing sector is the least concentrated. Appropriately enough, Vancouver and Calgary attain their highest S & T proportions in the resource and primary manufacturing sector, although Toronto and Montreal still dominate with 46 per cent of jobs. It is surprising, however, that less than 50 per cent of S & T jobs in the producer services sector are located in Toronto and Montreal. One would expect this sector to show the greatest degree of spatial concentration. Compared with Montreal, however, Toronto has a greater concentration of small S & T units, with a larger share of S & T units than labour force in all four sectors. Thus, if Canada is to be successful in generating additional R & D in both the production and service sectors, it is in the Toronto CMA that this will most likely be accomplished, because of the producer service development already in place.

Table 4.17 Canada: survey of scientific and technical activity, 1977: major metropolitan centres

	All sectors (%)	Producer services (%)	Technology-intensive manufacturing (%)	Other secondary manufacturing (%)	Resource and primary manufacturing (%)
A. Labour Force					
Toronto	34	29	34	30	25
Montreal	23	19	34	33	21
Vancouver	7	9	7	6	12
Ottawa-Hull	9	13	3	7	3
Calgary	5	7	1	1	10
Edmonton	8	6	2	1	9
Total CMAs*	(46,509) 100	(22,439) 100	(7,864) 100	(10,386) 100	(5,820) 100
B. Business Units					
Toronto	33	30	40	39	28
Montreal	18	13	20	21	17
Vancouver	10	12	10	9	11
Ottawa-Hull	8	10	5	6	4
Calgary	6	8	2	2	11
Edmonton	4	4	2	2	7
Total CMAs	(2,076) 100	(749) 100	(440) 100	(616) 100	(271) 100

* 22 CMAs.
Source: MOSST, 1978.

The level of S & T development in Toronto, however, falls short of the concentration of head office functions in the Toronto CMA (Tables 4.15 and 4.17). On a revenue basis, Toronto contains 48 per cent of Canada's non-financial head office activity while Montreal controls only 26 per cent (Semple and Smith, 1981). These shares, however, are a composite of foreign-controlled head offices—Toronto has 57 per cent and Montreal has 17 per cent—and those under domestic control—Toronto contains a more modest 39 per cent and Montreal 32 per cent. Furthermore, in Toronto, foreign head offices account for 54 per cent of the region's total, while the proportion for Montreal is only 30 per cent. It is apparent, therefore, that inflated expectations of Toronto's S & T activity would be generated if the whole of Toronto's office sector is used as the basis of comparison. More specifically, Toronto's foreign-controlled head offices are weaker generators of Canadian S & T employment. This is consistent with the facts, brought out earlier, that there are substantial international leakages from Canada of technological service requirements and the impacts made by foreign companies inevitably reduce the potential level of R & D activity in Canada. Furthermore, importing companies are predominantly foreign-owned and locationally concentrated in the Toronto region. The head office sector, however, is not the only foreign-controlled origin of reduced demand for producer services and limited S & T employment within corporations: the locational concentration of 45 per cent of foreign controlled manufacturing *production* occurs within 100 miles of the centre of Toronto and the behaviour of this sector, too, produces similar effects on S & T activity. While no complete and direct corroboration of the locational impact of foreign ownership on Toronto's S & T (and, therefore, R & D) employment can be made, the proposition advanced here is supported by a study by Shaw (1980) that reveals a low level of technological linkages with Canadian producer service firms for a sample of Toronto's manufacturing innovators. While this result applies to both foreign and domestic firms, Canadian-owned manufacturing is known to be a minor importer of technological services.

5. Canadian R & D policy

5.1 *Background*

Every industrial country, including Canada, has generated a set of policies intended to increase the chance of successful domestic industrial innovation by providing financial support for R & D. In Canada's case, attaining technological specializations is a substantial problem since Canada lies outside all major trading blocs. The programmed reductions in tariffs negotiated under GATT supposedly generate increased international competition which should encourage Canadian technology-intensive specializations. It must be recognized, however, that international competition occurs in an environment sustained by various forms of national industrial assistance, by the successful operation of non-tariff barriers against imports, and by the extraordinary marketing power of transnational corporations. These business realities, plus the competitive weakness of Canadian firms, are confronted by the Science Council of Canada and by various scholars who have stressed the need for:

1. larger Canadian core companies, in order to maintain the search for industrial innovation, to generate demands on smaller firms, and to sustain the penetration of domestic and foreign markets;
2. improved assistance for Canadian firms currently striving to become internationally competitive; and
3. recognition that current industrial R & D and product innovation has long-term significance and that only by maintaining industrial competence will Canada have a chance of retaining a place among the technologically strong economies.

In practice, the Science Council of Canada—a federally funded advisory unit initiating research, policy debates and policy development—has encouraged Canadian policy to move along a trajectory that is consistent with these ideas: national goals have been discussed and Canadian policy needs in relation to its industrial performance, and in comparison with other countries, have been outlined (Science Council of Canada, 1979, 1981). In response, economists within government departments have developed concepts that permit them to endorse similar policy initiatives. The Economic Council of Canada—a federally funded advisory body concerned with medium- and long-term problems of the economy—has also become a supporter and has shown itself willing to advise on the conditions under which tax concessions or direct grants could be justified, although it has not relinquished its advocacy of free trade.

Canada's technological policies have tended to be an outgrowth of industrial assistance programmes (always in search of clear policy goals) that were created as an adjunct to tariff protection. Those industrial policies, however, have never been consistent, since some programmes were concerned with regional development, particularly with industrialization outside the industrial core: while other programmes have been concerned with international competition, innovation and, in recent years, the survival of large firms in danger of failure.

Unclear goals, indefinite selection and award procedures, and uncertainty about the impacts of industrial assistance are an unfortunate inheritance for technological policy. Nevertheless, at the national and provincial levels, several policies and programmes have emerged. These fall into several categories.

Infrastructural improvements: enhancing the supply side. Government expenditures range from the development of 'research parks' to the establishment of centres concerned with the technology diffusion process. In the first case, the real estate development industry is well capable of initiating this type of project, but policy commitment is made visible to industry when provincial funds are used. In Ontario, infrastructural expenditures recently have been focused on six centres which are intended to enhance the technological knowledge of firms. Information networks will assist inter-firm contact. These technology centres operate on the understanding that industrial technology is often a public good but that since small firms have inadequate resources to search for relevant technical data, an information distribution system is a relevant and necessary public expenditure.

Manpower training, university participation in industrial research. The need to supply educational programmes that can train workers in highly skilled areas—in software, design, science, engineering and technology, and business—is recognized at all levels of government. In particular, the research training supplied by universities, and the experience that they can provide in the pursuit of industrial research, has been recognized as an important long-run policy commitment. University finance has been enduring a decade of reduction in real terms, however, and only recently have new programmes to expand research funding been developed and often these involve the financial participation of industry.

Industrial assistance: increasing the private sector demand for technology and private sector innovation. It is in this sphere that most criticisms of current policies and proposals are made by economists who favour a non-interventionist position (e.g. Palda, 1979). There are, nevertheless, strong theoretical (economic) and practical arguments in favour of incentives for firms to pursue innovation (Cohen *et al.*, 1984):

(a) firms have difficulty in appropriating all of the advantages from their R & D expenditures and, as a result, may underspend on R & D. Thus, the gains that derive from R & D in the wider local community are reduced;

(b) firms may also underspend on R & D because of the risk in this type of activity which individual firms are unable to spread over a large enough number of projects.

Cohen *et al.* (1984) go on to argue that since government can share the risks associated with R & D, government support in the form of direct loans and/or equity is warranted since the state may have lower (trans-action) costs than private financial institutions in coping with risk. The problems firms face in appropriating benefits of R & D expenditures, however, provide a cogent stimulus for government subsidies to raise the expected private economic rate of return, on socially desirable pro-jects, to that of the 'normal' rate of return on low-risk investments. Given these distinctions the case is made for both forms of instrument to be employed—and the case is made against tax-based R & D incentives like those adopted in Canada in recent years. Tax concessions are not fine-tuned to the needs of each firm on each project—the appropriability subsidy, for example, should vary from project to project, and in accord-ance with ownership status of the Canadian firm. Furthermore, in order for risk to be absorbed, direct loans, rather than indirect assistance, should be given. Cohen *et al.* acknowledge that administrative costs of discretionary assistance of these types will be high and suggest that for projects up to a given size limit the broad tax incentive approach may be appropriate. The limited availability of venture capital to small- and medium-sized firms is, however, a major weakness of this idea. Evidence collected from actual firms indicates that venture capital difficulties for small firms have been an impediment to growth (Steed, 1982). In addition, opinion seems to be growing that Canada's tax system, though providing greater incentives to undertake R & D than other Western economies (except Singapore), is poorly designed and complicated. Furthermore, incentives are often not of immediate benefit, and they provide 'windfall' gains (Brooks, 1984, 249). The evaluation of tax concessions as a policy instrument to encourage R & D can be undertaken more directly, however, using data that have been generated by recent expansions of the programme.

5.2 *Tax concessions as a component of Canadian R & D policy*

Enhanced tax provision for R & D activities was introduced in 1977, with several expansions since then, and, as noted above, resulted in an improvement in the level of R & D activity within the business sector over the past five years. An annual average rate of increase of 11 per cent in real business expenditures on R & D has been recorded in this period of recession! Business tax relief has been accompanied by a reduced

Table 4.18 Impact of tax incentives for R & D in Canada, 1980–81

Firm Size	Total intramural R & D expenditure*		Incremental R & D allowances†		Investment tax credit†	
(sales in $m.)	$m.	%	$m.	%	$m.	%
less than $10	149	7.6	6	3.3	4	4.8
$10–50	205	10.5	19	10.6	7	8.4
more than $50	1,606	81.9	155	86.1	72	86.7
Total	1,960	100.0	180	100.0	83	100.0
	No. firms	%	No. firms	%	No. firms	%
less than $10	515	46.7	60	20.1	54	20.3
$10–50	281	25.5	81	27.2	63	23.7
more than $50	306	27.8	157	52.7	149	56.0
Total	1,102	100.0	298	100.0	266	100.0

* Data for 1981 tax year.
† Data for 1980 tax year.
Source: Treasurer of Ontario, 1983, using data from Statistics Canada.

potential and commitment by government to act as a source of direct funding for R & D. This is a significant effect because the fiscal adjustments have assisted predominantly those businesses whose profitability and scale allowed them to undertake R & D prior to the introduction of the tax concessions. About half the large firms doing R & D claim incremental R & D and investment tax credits, whereas the proportion is as low as 10 per cent for small and 25 per cent for medium-sized firms. Given that the larger firms undertake larger projects, the bulk of the concessions go to these firms.

The sectoral impact of the new tax programmes is biased slightly in favour of the electrical and communications equipment industries (see Table 4.19) and against aircraft and other transportation equipment firms. Communications/electronics is the most effective sector from the standpoint of performer-funded R & D and in each of the three size ranges of firms (1981), firms in this sector out-performed the national aggregate and overall produced a performer-funded R & D sales ratio of $3.5: $100 sales (the national overall ratio being $0.9!). It is understandable, therefore, that this sector has generated substantial use of the new tax concessions. At the same time, however, the aircraft and other transportation equipment sector is unresponsive to tax incentives for R & D. There are two reasons; first, several large firms in this sector are non-taxable (mainly Crown Corporations) and second, many firms do little R & D in Canada (e.g. auto assembly). Non-taxable status applies also to the petroleum and related products sector (Petro Canada).

Although the improved tax conditions for industrial R & D seem to have had a positive effect, several offsetting factors limit the impact of the concessions. First, Canada has only a handful of very large firms—ten firms accounting for 35 per cent of R & D expenditures—which are best able to take advantage of the new arrangements. The tax benefits obtained by firms depend on their size and profits and large firms are favoured because their corporate tax rate is higher. The result is that, as of 1982, the net cost of incremental R & D expenditure for large firms could have been as low as 30 per cent (Table 4.20).

Second, there is evidence that many firms just 'do not appreciate the complexity of tax incentives, which few even among those eligible have used' (Steed, 1982, 140) and, therefore, there are serious doubts about the effectiveness of the way policy is delivered.

Third, the definition of R & D by 'Revenue Canada' is heavily biased in favour of scientific research; and, according to Steed, past *interpretations* have minimized the acceptance of expenditures concerned with 'exploratory development work, trial production and engineering follow-through'

Table 4.19 R & D Expenditures and tax incentive claims by industry group, Canada, 1980

Taxable firms	Total expenditures (%)	Invest. tax credits (%)	Special allowance (%)	Total (%)
Elect./comm. equip	21.3	30.2	27.2	28.3
Aircraft and other transport equipment	13.9	1.7	1.7	1.7
Petroleum products, oil and gas wells	15.8	9.6	17.8	14.7
Chemicals	8.5	6.6	4.7	5.4
Other manufacturing	26.0	25.1	18.1	20.8
Other industries	14.6	26.7	30.4	29.0
Total	100.0	100.0	100.0	100.0

Source: Revenue Canda.

Table 4.20 Net R & D costs to Canadian firms per $100 of expenditure

Firm size for tax purposes	Ongoing R & D expenditure	Incremental research expenditure
Small	$56	$44
Large	$51	$30

Source: Ontario, Economic Research Branch.

(Steed, 1982, 143). The consequence is that the supposedly favourable tax treatment of R & D is poorly matched to the full innovation sequence.

The ineffectiveness of tax concessions as an R & D stimulus for small (and some large) firms that have become non-taxable or have insufficient cash-flow to initiate R & D was recognized in the 1983 tax year. By means of Scientific Research Tax Credits (SRTCs), firms were permitted to exchange unused tax concessions for equity investment by individuals or institutions with debt reduction permitting R & D expenditures. This arrangement complements similar programmes to encourage equity investment such as the Small Business Development Corporations (SBDC) programme of Ontario. Despite this innovation, Canada's fiscal policies to increase R & D are likely to be frustrated by the current recession.

5.3 *Direct government funding of R & D*

While R & D assistance through the tax system in Canada is reported to compare favourably with other countries (Brooks, 1984), direct government funding remains inordinately low (Table 4.21).

Nevertheless, in practical terms, the advantages of direct grants or loans are:

1. non-taxable firms receive support, particularly small technology-intensive companies without sufficient cash-flow and large companies that have lost taxable status, and the support is delivered before expenses are incurred;
2. assistance may be directed to industries, firms and projects that conform to the country's developmental goals: parenthetically, Canada's difficulty in setting those goals may have promoted the tax-based route of assistance; and
3. a successful 'track-record' of the launching of profitable new products by Canadian firms as a result of direct subsidization exists—though repeated 'write-offs' of the debts of several crown corporations, including

Canadair and de Havilland, indicate that direct assistance for R & D in Canada has involved major problems of accountability.

An additional and significant problem encountered in the past in assessing Canadian direct grant support of R & D, innovation and technology transfer has been the relatively large number of related but unconnected programmes and the paucity of data on funds that have been allocated and on the decision systems employed. Over recent years, however, programmes like EDP have amalgamated a variety of formerly separate arrangements (Table 4.21). The current range of direct assistance includes cost-sharing of selected projects (salary of unemployed engineers, scientists or technicians), co-operative research between firms and NRC laboratories to effect technology transfer, assistance with innovation in specific sectors, technology development for export in defence firms, contracting-out by government to industry, funding of unsolicited proposals for scientific work and government procurement of technology-intensive goods.

Table 4.21 Expenditures by Canada's major programmes of direct federal assistance for R & D and innovation, 1979-80 and 1981-2 ($m.)

	1979–80	1981–82
PILP	6.0	15.0
IRAP	19.5	23.9
DIPP	38.6*	38.6*
EDP	84.6	119.2
Total	148.7	196.7

* Average 1969–79.
Notes: PILP: Programme for Industrial/Laboratory Projects
    IRAP: Industrial Research Assistance Programme
    DIPP: Defence Industry Productivity Programme
    EDP: Enterprise Development Programme
The first two programmes are administered by the National Research Council, the second two were functions of the Ministry of Industry, Trade and Commerce.
Source: Economic Council (1983, 144-51).

Canadian government assistance to industry extends through a wider network of programmes than listed in Table 4.21, but the Federal Business Development Bank (a lender of last resort) and the Export Development Corporation concerned with financing exports, for example, are the Canadian version of similar schemes in most OECD countries and are excluded from the programmes that address R & D or innovation directly. Although various sources, all supposedly using the same data, generate different tables (see Abonyi and Atkinson, 1983, 104), Table 4.21 has been abstracted from information compiled by the Economic Council (1983). The $200 million current assistance for R & D (a figure generally corroborated in Statistics Canada, 1980, 11) may be compared with the $263 million of Federal Tax Credits (1980) that were used to lever substantial private R & D expenditures. Estimates or SRTC costs range as high as $1 billion in tax credits for 1984.

The Economic Council's (1983) evaluation of programmes designed to encourage R & D concludes that while they have 'a legitimate place . . . their efficiency depends entirely upon how judiciously and effectively they are employed. It is apparent that there is a good deal of scope for improvement' (Economic Council, 1983). Their argument is focused on how need and impact are appraised by agencies responsible for assisting projects. Looking at the programmes from the standpoint of the operations of firms, however, generates other considerations. It is apparent from various enquiries that the goals and mandates of government programmes need to be expanded past the limiting R & D definition employed most commonly by economists and statistical agencies such as Statistics Canada and the OECD. It was argued earlier that while industrial innovation depends on the level of R & D expenditure, other knowledge-based and skill-intensive activities are required to implement innovation. Engineering, design, production start-up and marketing are of particular significance. Canada has neglected to support adequately these later stages of product innovation. Although government programmes are directed towards research, Ondrack's evidence for the industrial machinery sector indicates that smaller Canadian-owned companies emphasize product development and that marketing assistance is needed for small firms if they are to export (Ondrack, 1975). His later research confirms that R & D in the machinery industry is 'far more applied than what the assessors were prepared to accept and therefore, the R & D support programs were of limited relevance' (Ondrack, 1980, iv). Nevertheless greater understanding of the later phases of innovation has penetrated policy circles over the past five to ten years and in the case of the Ontario Research Foundation, which undertakes contract technological work (testing, analytical services, design and development, engineering, facility layout and other productivity improvement advice), much of the financial support is obtained by individual firms through government programmes. The National Research Council (NRC) sponsors a Technical Information Service through ORF, thus providing advice on possible sources of financial assistance.

5.4 *Provincial government R & D funding*

The Canadian provincial governments fund only a minor share of Canadian R & D. In 1982, for example, funding from the provinces was only 5.6 per cent of gross expenditures on R & D while the federal government contributed 35 per cent. Nevertheless, provincial expenditures have maintained their real share of R & D in part because of a deliberate attempt by Western provincial governments to increase R & D (at twice the average funding level). Of the total provincial effort, about one-third is made through provincial research organizations (PROs) that are distinct from research units in government departments. About one-third of their expenditures are concerned with experimental development and just over 20 per cent with scientific research; at the same time, one-third goes to manufacturing firms and a quarter to primary industries. There has, however, been little effective interaction between the federal and provincial jurisdictions, and Jenkin's conclusion is that weak federal leadership and a growing experience of provincial agencies in interacting with firms means that the Canadian R & D policy is at a crossroads if structural change in the economy is to be addressed creatively (Jenkin, 1983).

5.5 *R & D policy and foreign ownership*

Foreign direct investment has been addressed explicitly through policy measures since the introduction of the Foreign Investment Review Agency a decade ago. Its mandate, however, deals with take-overs and other new investments; although it has negotiated the conditions under which approvals have been given, it has not had much impact on the basic issue of R & D by foreign firms that were already part of the Canadian industrial structure.

Foreign firms make investment and employment decisions in accordance with priorities that reflect the multinational scope of their operations. R & D is typically a headquarters function for reasons of organization (control), strategy (appropriability of the gains of innovation) and finance (economies of scale in R & D). For the most part, R & D for Canadian production is undertaken by multinationals in a variety of US cities and, as noted earlier, R & D actually performed in Canada is usually concerned with adapting foreign technology rather than generating a Canadian-based, knowledge-intensive capability. Other reasons for R & D performed in Canada include benefiting from Canadian government grants (which are probably 'windfalls') and pursuing world product mandates in accordance with exceptional company policies that align a portion of corporate growth and specialization with Canadian industrial development.

Canada stands to gain substantially from world product mandating by multinationals and the Science Council and the Province of Ontario (Ontario Ministry of Industry and Tourism, 1980) have both recognized this mandating as an important mechanism whereby Canada may increase technology-intensive exports, domestic sourcing, product innovation, and R & D, and may reduce imports by MNCs. In Ontario it has been argued that falling Canadian tariffs should induce a shift from 'the traditional branch plant form of operation' to 'specialized missions within the worldwide enterprise' (Ontario, Ministry of Industry and Tourism, 1980, 2-3). While we should be cautious when specifying the nature of these 'missions', at least the view is more realistic than the one often cited, that falling tariffs will force some Canadian (domestic) firms to be internationally competitive.

The policy initiatives that have been suggested in Ontario focus on consultations to encourage more Canadian affiliates to obtain 'specialized missions', by generating assistance for Canadian managers to develop their cases wtih headquarters and seeking export support, government procurement, research and development subsidies (called incentives) and performance incentives and grants for affiliates. Recent reports suggest that declining tariffs are acting as a catalyst for the revision by multinational firms of their Canadian operations. One result is that Canadian subsidiaries are being retained on the understanding that product innovations (for world markets) will reach the production stage relying on Canadian research, engineering, and design. This form of international division of labour would be a complete transformation of the 'miniature replica' branch plant of the past. Declining tariffs are also precipitating less positive forms of change by foreign subsidiaries: disinvestment strategies reduce manufacturing plants to warehouse status as US firms cut direct participation in the Canadian manufacturing sector in a programme of North American rationalization of investment in plants. As argued elsewhere (Britton, 1978), if *production* facilities in a number of industries in the USA are shifting to the south (in accordance with lower labour costs and little unionization), and can supply north-eastern US cities at lower cost, then declining Canadian tariffs mean that the exchange value of the Canadian dollar and the Canadian wage rate are the major factors preserving Canadian production jobs in currently less efficient plants. Increased scale of production, exports, and efficiency might, of course, result as tariffs decline if US firms preserve Canadian plants as part of their North American networks, but as long as innovation-directed activities are located at headquarters in the USA the technological learning opportunities for Canadian firms are virtually zero.

6. Conclusions

By international standards Canada has significant deficiencies in R & D expenditures. Much of the gap is attributable to insufficient expenditure by business in Canada. The response of government has been vacillation in developing a policy for the country and rather than making up the shortfall in industrial funds, resources have been directed to government laboratories and tax concessions have been created in order to encourage R & D. Canada's lacklustre trade performance in goods with substantial technological content has probably been one stimulus of these actions but there have been other concerns, too. Tariff protection of Canadian firms is declining and it is obvious that they need assistance in order to generate an innovative base for more effective international competition. The increasing pressure of imports of goods from Third World producers has also made it clear that Canada has little temporal leeway in which to establish industrial specializations that will offset imports and will expand exports.

Canada, then, has taken some steps to diversify its comparative advantage from one dominated by its resources (a richly endowed but a high cost and increasingly marginal competitive export base). Nevertheless, labour cost advantages (compared only with the USA) and reduced tariff protection provide an inadequate advantage for manufacturing firms and the direct and indirect jobs that they have generated in the past. The restructuring of the US economy over the past decade, in particular, has shown the vulnerability of Canadian manufacturing to the impact of capital mobility and functional reorganization within multinational corporations. Foreign ownership is responsible for much of the weakness in Canada's trade and industrial structure and this is illustrated by both the technology imports made by foreign subsidiaries and by their low comparative expenditures on R & D in Canada. The lack of an industrial-technological strategy for Canada is revealed starkly by the lack of resolve to tackle the issue of differential technology assistance on an ownership basis. Canada's choice of expanded tax concessions to firms undertaking R & D is thus a policy that neither assists emerging or potentially innovative

Canadian firms, nor respects the theoretical arguments that indicate that specific loans and grants should be employed to offset the problems of uncertainty and appropriability encountered by firms undertaking or planning R & D expenditures. Furthermore, there seems to be a basic policy resistance to assisting firms in the latter stages of the innovation process.

Canada has some claim to excellence in technologies related to the geographic size, resources and environment of the country, but the economic size of the Canadian market has made the development of internationally significant and lasting industrial specializations based on these technologies elusive. The dominance of foreign direct investment in the technology-intensive industries makes this problem worse because Canadian companies are thwarted in their search for market-based growth, and foreign (mainly US) companies have proved to have minimal interest in using Canada as a base for world market ventures relying on Canadian-located R & D. World product mandating, although adopted by a minority of multinationals in Canada, remains the fragile hope that is held for more technological investment by foreign firms in Canada.

Toronto, Canada's major metropolitan centre, is the largest locational concentration of R & D activity. Nevertheless, its metropolitan functions —especially corporate headquarters—have not generated a comparable concentration of technological activity. To a large degree, this situation is a result of the large proportion of its 'head offices' that are only regional offices of US firms. Functionally they are 'hollow' by comparison with their parent offices in US cities. Toronto is also the centre for US production in Canada (especially in secondary manufacturing) and thus it is the focus of imports of parts, assemblies and services from the USA. The implication is that Toronto is also the Canadian focus of opportunity costs of industrial development that stem from the reduced market thresholds that inhibit the birth and growth of specialist services and manufacturing firms and their expected multiplier effects (Britton and Gilmour, 1978). Nevertheless, there seems to be a metropolitan specialization emerging in Toronto that depends on numerous firms in the producer services and their existence and development is a positive sign of growing industrial complexity and restructuring.

Many of Toronto's manufacturing firms have been assisted by a variety of grant and loan programmes that have been generated over the past two decades. Most of these programmes, however, have presented bureaucratic problems for firms and they have been too numerous and too small. Recently, federal government programmes have been reduced in number through amalgamation, and regional and industrial ministries have been integrated. The regional offices of the Department of Regional and Industrial Expansion, with a mandate to support industrial innovation, should provide a more sensitive delivery of subsidies to small enterprises than has been true in the past. At the same time, the recession which has hastened employment decline in manufacturing and outright plant closures has also stimulated local authorities to consider assisting small industrial firms. Realism suggests that lasting job replacement will only derive from the birth and expansion of innovation-dependent enterprises and in Toronto and other industrial centres new forms of public policy are emerging to assist industrial incubation and technology diffusion.

Notes

1. In real terms industrial R & D expanded by 112 per cent over the period.
2. Localization was evaluated using NSEM location quotients greater than 1.5 with a base of the labour force.
3. Log NSEM growth $= -1.91 + 1.13$ log Labour Force
$$(0.05)$$
$r^2 = 0.97\ n = 21$
4. Log S & T $= -1.82 + 1.22$ log NSEM
$$(0.08)$$
$r^2 = 0.93\ n = 22$
5. Scientists, engineers and technicians comprise 70 per cent of the labour force total in the survey.

References

Abonyi, A. and Atkinson, M. M. (1983). 'Technological Innovation and Industrial Policy: Canada in an International Contest', in Michael M. Atkinson and Marsha A. Chandler (eds), *The Politics of Canadian Public Policy* (Toronto, University of Toronto Press).

Britton, John N. H. (1974). 'Environmental Adaptation of Industrial Plants: Service Linkages, Locational Environment and Organization', in F. E. Ian Hamilton (ed.), *Spatial Perspectives on Industrial Organization and Decision Making* (London, Wiley), pp. 363–90.

Britton, John N. H. (1976). 'The Influence of Corporate Organization and Ownership on the Linkage of Industrial Plants: A Canadian Enquiry', *Economic Geography*, 52, 311–24.

Britton, John N. H. (1978). 'Locational Perspectives on Free Trade for Canada', *Canadian Public Policy*, 4, 8–12.

Britton, John N. H. and Gilmour, James M. (1978), *The Weakest Link* (Ottawa; Science Council of Canada, Background Study 43).

Brooks, Neil (1984). Review of Donald G. McFetridge and Jacek P. Warda, *Canadian R & D Incentives: The Adequacy and Impact* (Toronto, Canadian Tax Foundation, 1983), *Canadian Public Policy*, 10, 248–9.

Chand, U. K. Ranga (1980). *Characteristics of Research and Development Performing Firms in Canadian Manufacturing* (Ottawa, Policy Research Group, Industry Branch, Ministry of State for Science and Technology).

Cohen, J. S., Rubin, J., and Saunders, R. S. (1984). 'Chasing the Bandwagon: Government Policy for the Electronics Industry', *Canadian Public Policy*, **10**, 2, 534.

Daly, Donald J. (1979). 'Weak Links in the "Weakest Link"', *Canadian Public Policy*, **5**, 307–17.

de Melto, Dennis, McMullen, Kathryn E., and Wills, Russel M. (1980). *Preliminary Report: Innovation and Technological Change in Five Canadian Industries* (Discussion paper 176, Ottawa, Economic Council of Canada).

Economic Council of Canada, 1983. *The Bottom Line: Technology, Trade and Income Growth* (Ottawa, Economic Council of Canada).

Frankl, R. (1979). *A Cross Section Analysis of Research and Development Intensity in Canadian Industries with Particular Reference to Foreign Control* (Ottawa, Industry, Trade and Commerce, Economic Policy and Analysis).

Hayter, Roger (1982a). 'Truncation, the International Firm and Regional Policy', *Area*, **14**, 277–82.

Hayter, Roger (1982b). 'Research and Development in the Canadian Forest Product Sector—Another Weak Link?', *The Canadian Geographer*, **26**, 256–63.

Hewitt, G. K. (1981). *R & D in Selected Canadian Industries. The Effects of Government Grants and Foreign Ownership* (Ottawa, Industry, Trade and Commerce, Technology Branch).

Jenkin, Michael (1983). *The Challenge of Diversity* (Ottawa, Science Council of Canada, Background Study 50).

Malecki, Edward J. (1983). 'Technology and Regional Development: A Survey', *International Regional Science Review*, **8**, 89–126.

Ondrack, D. A. (1975). *Foreign Ownership and Technological Innovation in Canada: A Study of the Industrial Machinery Sector of Industry* (Ottawa, Industry, Trade and Commerce, Technology Branch).

Ondrack, D. A. (1980). *Innovation and Performance of Small and Medium Firms: A Re-analysis of Data on a Sample of Nineteen Small and Medium Firms in the Machinery Industry* (Ottawa, Industry, Trade and Commerce, Technology Branch).

Ontario Ministry of Industry and Tourism, 1980. *The Report of the Advisory Committee on Global Product Mandating.*

Palda, S. (1979). *The Science Council's Weakest Link: A Critique of the Science Council's Technocratic Industrial Strategy for Canada* (Vancouver, The Fraser Institute).

Rugman, Alan M. (1981). 'Research and Development by Multinational and Domestic Firms in Canada', *Canadian Public Policy*, **7**, 604–16.

Safarian, A. E. (1979). 'Foreign Ownership and Industrial Behaviour: A Comment on "The Weakest Link"', *Canadian Public Policy*, **5**, 318–35.

Science Council of Canada (1979). *Forging the Links: A Technology Policy for Canada* (Ottawa, Report 29).

Science Council of Canada (1981). *Hard Times, Hard Choices: Technology*

114   John N. H. Britton

and the Balance of Payments (Ottawa, Industrial Policies Committee).

Semple, R. Keith and Randy W. Smith (1981). 'Metropolitan Dominance and Foreign Ownership in the Canadian Urban System', *The Canadian Geographer,* 25, 4–26.

Shaw, Alison (1980). *Manufacturing in Toronto and the Innovation Process* MA Research Paper, Department of Geography, University of Toronto.

Statistics Canada (1980). *Annual Review of Science Statistics,* Cat. No. 13-212.

Statistics Canada (1983). *Standard Industrial R & D Tables 1963–1983,* Cat. No. 13-X-201 (Ottawa, Science Statistics Centre).

Steed, Guy P. F. (1982). *Threshold Firms: Backing Canada's Winners* (Ottawa, Science Council of Canada, Background Study 48).

Treasurer of Ontario (1983). *Ontario Budget Paper: R & D and Economic Development in Ontario—A Discussion Paper.*

5 Public Sector Research and Development and Regional Economic Performance in the United States*

E. J. MALECKI
Department of Geography, University of Florida at Gainesville

1. Introduction

Government R & D fits uneasily into research on technological change and regional economic development. On the one hand, public sector R & D falls into the category of national policies with *implicit* regional impacts; for example, a set of policies that have principally national objectives unrelated to regional conditions, but effects that often run directly counter to the aims of *explicit* regional policies (Glickman, 1980). On the other hand, public sector R & D is part of a total regional R & D effort. Consequently, since the regional incidence of total R & D is often assumed to be correlated with regional growth, there is an implication that similar beneficial regional impacts will ensue from the location of both private *and* public sector R & D. As Buswell (1983, 18) mentions when considering R & D generally, 'there remains something more than a lingering suspicion that the two—R & D and regional development—are related.' However, the causal links are more difficult to establish, in part because of the lengthy time horizons between some R & D, such as the basic research often performed by public institutions, and its downstream effects on industrial innovation, employment and other indicators of regional economic performance.

This chapter addresses both roles played by public sector R & D: first, as part of R & D policies at a national level; and second, the role of R & D and its location as a vehicle for explicit regional change. The following section focuses on government R & D as a portion of total R & D activity. It is demonstrated that the sectoral orientation of government R & D generates a different set of "expected" effects on regional economic specialization from that of R & D in the private sector. The paper then turns to the topic of regional impacts, for which a larger number of studies are available for comparative purposes, although they have tended to be largely impressionistic rather than systematic in their appraisal of the regional effects of R & D activity. The final section provides a synthesis

* This research was supported by the US National Science Foundation, Division of Policy Research and Analysis. I wish to thank Roger Roberge for comments on an earlier version of this chapter.

of current knowledge on the interrelationship between R & D in the public sector and regional economic performance. Although most of the empirical examples are drawn from the American experience, many of the generalizations are applicable to other advanced nations, including Canada, France, the United Kingdom, and West Germany, as Van Duijn and Lambooy (1982) have suggested.

2. R & D in the public sector

Most conceptual treatments of government R & D consider it merely a supplement to R & D funded by the private sector, and then only in terms of certain specific functions. For example, basic research for fundamental sciences and R & D related to 'public goods,' such as defense, medicine, and energy, whose benefits accrue to society as a whole, are two such generic categories (Fusfeld, Langlois, and Nelson, 1981). R & D, however, is a rather 'lumpy' activity in spatial terms, performed best in many cases by large research teams in specific locations. The sectoral, organizational, and locational settings of such R & D greatly affect its influence on other economic activities within a region. Sectorally, government R & D in several major countries (including the USA, UK, and France) is heavily oriented toward one 'public good'—defense and military applications (OECD, 1981). Moreover, large firms are the major performers of government-funded, as well as privately funded, R & D in all the major OECD countries. The majority of R & D is performed by the business sector in most such countries (Table 5.1), and concern is

Table 5.1 Distribution of R & D expenditure by sector of performance in leading OECD countries, 1977

Country	Business enterprises	Government	Higher education	Private non-profit
United States	66.8	15.2	14.7	3.3
Japan	57.8	12.1	27.7	2.4
Germany	65.0	16.1	18.6	0.4
France	60.3	22.8	15.5	1.4
United Kingdom*	60.3	25.9	10.4	3.3
Netherlands	51.7	20.8	25.0	2.5
Canada	41.9	30.5	26.8	0.8

* 1975–6.
Source: OECD, 1981, 49.

expressed about the concentration of contracts in very large enterprises (OECD, 1981, 70–1). Government R & D can be even more concentrated than private R & D activity. In the United States, for example, 82.6 per-cent of all government R & D performed by firms in 1979 went to those with over 75,000 employees. By comparison, industry-funded R & D by such firms accounted for 67.6 percent of total industrial funding.

Government R & D is also sectorally and locationally biased, much like that of other large, but private sector organizations. A very small number of industrial categories receive virtually all the publicly supported R & D. Even at first glance, the sectoral allocations make plain the domin-ance of a military concern in the government R & D picture in the USA. Of $12.5 billion of federal R & D performed by industrial firms in 1979, 75.4 percent was to firms in just two industrial sectors: aircraft and missiles, and electrical equipment. In France, the situation is similar (Norman, 1979). Even as defense declined below 45 percent of all US federal R & D temporarily during the late 1970s, the growth area of energy-related R & D was often directed toward the same large corpora-tions that had diversified to take advantage of these large, technologically advanced programs. Several firms have focused their business on defense, space exploration, and energy, where government purchases of products as well as support of R & D are likely (Gansler, 1980; Pavitt, 1979).

The regional distribution of R & D in the United States is strongly influenced by the mix of performers of R & D. Private sector R & D consists principally of product development rather than new processes. Scherer (1982) estimates that 73.8 percent of all American industrial R & D went to either consumer products or capital goods to be used in other industries as process innovations. Government-funded R & D is even more heavily oriented toward product innovation (Kochanowski and Hertzfeld, 1981). The effects of R & D on regional economic per-formance, then, are much more complex to trace, since they depend in large part on the input–output linkages between R & D-performing and R & D-using sectors (Lee, 1983; Scherer, 1982). Direct effects of public sector R & D on regional productivity growth or output growth will be especially difficult to identify.

Any regional effects also must take into account R & D performed within government facilities and by universities, colleges and non-profit-making institutions, as well as that performed by industry. Table 5.2 shows the performers of American public sector R & D in Fiscal Year 1981, broken down by the agency source of R & D funding. Defense and space R & D, which together account for two-thirds of all govern-ment R & D spending are performed mainly by industrial contractors,

Table 5.2 Performers of US federal R & D by agency, fiscal year 1981 (by percentage of total R & D)

Agency	Intramural	Industrial firms	Universities and colleges	Other non-profit institutions	State and local governments	FFRDCs*	Total† ($m)
Defense	25.9	65.6	3.4	0.8	0.0	3.7	16,508.6
NASA	24.9	64.5	3.4	2.8	0.0	2.2	5,406.6
Energy	9.2	27.1	6.1	1.3	0.1	55.7	4,918.2
HHS	22.2	4.1	55.6	14.3	1.9	1.1	3,927.1
NSF	11.1	3.1	73.0	4.8	0.2	7.8	961.6
Agriculture	66.1	1.1	31.4	0.3	0.5	0.0	774.0
Other agencies	46.3	17.1	12.0	6.5	5.7	7.5	2,421,3
Total federal	25.0	46.6	12.8	3.2	0.6	10.8	34,917.4

* Federally-funded R & D centers.
† Totals ($m) exclude R & D performed in foreign locations, most of which is for the Defense Department or NASA.
Source: National Science Foundation (1982), *Federal Funds for Research and Development, Fiscal Years 1981, 1982, and 1983 (Detailed Statistical Tables)* (NSF 82-326), Washington, DC: National Science Foundation, pp. 26–7.

and in government 'intramural' facilities. Energy R & D, which grew rapidly through the 1970s but is now in decline, is conducted largely in government laboratories (called Federally Funded Research and Development Centers or FFRDCs), many of them located in isolated areas. Biomedical R & D sponsored by the Department of Health and Human Services and basic scientific research funded by the National Science Foundation are both performed mainly in universities and medical centers, while agriculture R & D is performed in small government facilities.

A sharp contrast in regional orientation results from this distribution of performers. The R & D of the 'big science' agencies (especially Defense and NASA) is determined by the corporate performers, concentrated at the small number of sites utilized by the relatively few sectors and individual firms involved. Government facilities, whether intramural or FFRDCs, are 'lumpy' in distribution, whereby a few large defense and energy laboratories or clusters of facilities account for much of the overall pattern. By contrast, agriculture R & D facilities are scattered fairly evenly around the country, and university research is also rather widely dispersed. The small portion of total public R & D that is oriented toward widespread distribution, however, is not able greatly to alter the geographical patterns. Table 5.3 indicates a sharp polarity among American regions in the relative concentration of public sector R & D. The location quotients in the table are the ratio between per capita R & D in a region and the national per capita figure, thereby controlling for regional population. This standardized basis for the examination of government expenditures reveals that, in the USA, four of the nine census regions have above-average levels of total federal R & D (location quotient greater than 1.0). The other five regions, on the other hand, receive little public R & D expenditure in any performer category.

These patterns have changed little since 1965, and any shifts which have occurred have tended to reinforce rather than alter existing patterns. In each region, some performer, or combination of performers, can be seen to have contributed to these changes. In New England, industrial R & D doubled its regional share, while all other performers remained near or above national norms, or improved slightly. New England is highest ranked for academic and non-profit institutions. Location quotients in the Pacific region declined for all performers but one (universities and colleges). However, this region remains the only one in the country to be disproportionately high for *all* performer categories, and is highest ranked for industrial firms. The Mountain region is also above average for all but one performer (non-profit institutions), but its values remain closer to 1.0, except for FFRDCs, in which the Mountain region dominates.

Table 5.3 Location quotients for US federal R & D by performer, fiscal year 1979

Region	Total federal R & D	Performers					
		Federal intramural	Industrial firms	Universities and colleges	Other non-profit institutions	FFRDCs	Other
New England	1.76	0.97	2.05	2.40	4.23	0.98	1.34
Middle Atlantic	0.69	0.48	0.58	1.01	1.12	1.07	1.28
East North Central	0.41	0.47	0.25	0.75	0.61	0.47	0.38
West North Central	0.60	0.20	0.89	0.82	0.60	0.06	1.10
South Atlantic	1.26	2.65	0.84	0.99	0.87	0.22	0.86
East South Central	0.75	1.06	0.59	0.55	0.35	1.04	0.27
West South Central	0.50	0.46	0.63	0.60	0.29	0.0	0.62
Mountain	1.61	1.33	1.07	1.13	0.85	5.25	2.93
Pacific	2.00	1.05	2.74	1.32	1.28	2.25	1.40

Source: Calculated from data in National Science Foundation (1980), *Federal Funds for Research and Development, Fiscal Years 1979, 1980, and 1981 (Detailed Statistical Tables)* (NSF 80-318), Washington, DC: National Science Foundation, pp. 144–54.

The South Atlantic region, which includes the Washington, DC, area, dominates in intramural R & D (the only category in which this region is above average), notwithstanding its slight improvement since 1965 in three categories (firms, universities, and non-profit institutions). In the Middle Atlantic region, a declining proportion of national R & D was due to falls in intramural and industrial R & D, despite expansion in the FFRDC, non-profit and other performer categories. The East North Central and West South Central regions have consistently experienced below-average location quotients in all categories, while these two regions also have the lowest overall location quotients, neither higher than 0.5.

The broad patterns of public sector R & D overlap considerably with that of private sector activity, especially in the locations of federal R & D performed by industrial firms. The other categories of R & D are relatively small and do little to affect the observed distribution. Although the Pacific region and all three regions in the north-eastern part of the country (New England, Middle Atlantic, and East North Central) have above-average levels of industrial R & D, only two regions—New England and Pacific—have above-average R & D in both the public and private sectors. Only in these two regions is there a roughly equal, high level of both public and private sector R & D. Thus, an important impact of government R & D activity comes from the agglomeration of technical employees. Firms engaged in R & D are attracted to such locations, since technical workers are found there in abundance. Although the processes are not well understood, employment generation via new firm formation, spin-offs, and corporate innovative activity seems to be most likely in existing loci of R & D and related activity (Malecki, 1981; Van Duijn and Lambooy, 1982). Consequently, the US North-east and West Coast continue to attract and retain a disproportionate amount of industrial R & D activity.[1]

The formation of agglomerations of R & D, private as well as public, appears to be the critical input for the alteration of regional competitiveness. Government R & D, whether at agency laboratories or contracted to firms and universities, is alone probably not sufficient to spawn wide-ranging technology-based growth. The technological activities of firms are also required, ranging from the commercial R & D activities of large firms to the exploitation of opportunities by new firms. Research by Cooper (1971) in Silicon Valley showed that government labs were incubators for few new firms, perhaps because of their lack of market orientation or because of their employees' low level of interaction with private sector counterparts. In addition, much government R & D is in defense-related fields, into which there is relatively little opportunity for new, small firms to enter (Gansler, 1980).

The mutual attraction of private and public R & D can be examined statistically by attempting to account for the location of industrial R & D activity, as influenced by government R & D spending patterns. Table 5.4 shows the results of a two-stage regression model of the number of industrial R & D laboratories in the USA by state in 1982 per thousand of the population. The first stage accounts for and removes the effect of federal funding of R & D by industrial firms (per thousand of the population) on the distribution of industrial R & D labs—a necessary step since data separating the two activities are not available. The second stage accounts for the separate effects of federal R & D at universities, agency laboratories and at FFRDCs. In this stage of the model, the only

Table 5.4 Two-stage least-squares results of influence on industrial R & D location by state, 1980

Stage 1	Intercept	25.93	
	Federally-funded industrial R & D per 1,000 population	0.199 (3.466)	$R^2 = 0.200$
Stage 2	Intercept	−13.489	
	Federal R & D by universities per 1,000 population	0.907 (3.200)	
	Federal intramural R & D per 1,000 population	−0.0030 (0.005)	
	FFRDC R & D per 1,000 population	−0.048 (1.064)	$R^2 = 0.232$

Notes: Dependent variable is number of industrial R & D labs per 1,000 population.
Federal R & D data are for FY 1979.
$N = 50$ (The District of Columbia is omitted from this analysis).
Values in parentheses are *t*-values.

component of federally funded R & D with a significant coefficient is that performed at universities (Table 5.4). These results at the state level suggest that research universities are the principal attraction related to government policy that affects industrial R & D location, and the only statistically significant one. In this geographical cross-sectional analysis, the large government R & D facilities seem to hold relatively little attraction for industrial R & D.

3. Localized regional impacts

The distribution of government R & D in the United States has had important impacts on the economies of some regions. However, it has taken a combination of factors, many of them related to the private sector, to generate the high-technology regional economy that is the goal of many regions. The results of the previous section suggest that university R & D, although only a small part of the total outlays from government, has a disproportionately beneficial effect on the regions in which it is concentrated. Other areas may benefit from certain large-scale government R & D in specialized facilities, such as the NASA complex and its spin-offs near Cape Canaveral (Holman, 1974), or the growth of Ottawa's technological complex in Canada (Steed and DeGenova, 1983). These facilities and their impacts are far from commonplace, however, and cannot be relied upon by most regions for regional economic vitality.

The academic sector is a small but important part of a nation's R & D picture (Table 5.1). University researchers perform much of all basic research as well as train future researchers for work in government, industrial, and academic settings. However, despite the large number of academic institutions in the USA (over 500 perform some R & D), relatively few of them attract the bulk of research funds. The leading American universities and colleges in R & D have been a fairly stable set for several years. Federal R & D funds account for just over 66 percent of all academic R & D, and are the largest source of R & D funds in nearly all academic institutions. The other principal source is state and local government support, which is especially important in the case of state-supported universities. The regional pattern of federal support of university R & D is remarkably similar to that of total federal R & D. The New England, Mountain, and Pacific regions stand out as well above the national level; the Middle Atlantic region is slightly above average.

To a great extent, the geographical patterns reflect the locations of the small number of major research universities. The top twenty-five universities have consistently received over 45 percent of all federal academic R & D. These twenty-five institutions are found in only fourteen states, but only five states have two or more leading universities; these are California (six), New York (four), while Illinois, Massachusetts, and Pennsylvania each have two. This represents an increased geographical concentration since 1965, when sixteen states were represented in the top twenty-five. (California and New York are the states that have gained.) The geographical pattern of public R & D at universities illustrates the cumulative concentration of research strength. In Fiscal Year 1979, five states received over 45 percent of the US total (California, New York, Maryland, Massachusetts, and Texas), while 82 percent went to the top twenty states. Defense and NASA R & D funds to universities are even more concentrated in a few states. In general, the combination of two or more research-oriented universities appears to ensure the attraction of large amounts of federal academic R & D to a region. The strongest areas of publicly supported research, California and Massachusetts, have been the origins of much recent activity in technology-based regional growth in biotechnology, microelectronics, and computer technology. The concentration of research universities, mainly in the North-east and the Pacific Coast, also coincides with major locations of federal R & D for defense applications in aerospace, electronics, and computers.

A larger direct impact of public R & D on regional performance might be expected from government R & D facilities, such as agency laboratories and FFRDCs which together comprise the bulk of non-industrial federal R & D effort. These establishments receive nearly three times the aggregate R & D funding of universities, and they are in relatively few locations, compared to academic research. FFRDCs are found in only twenty-one states and, although intramural R & D is found in every state, 34.8 percent of it is concentrated in the vicinity of Washington, DC—in Maryland, Virginia, and the District of Columbia. When California is added, 48 percent of all intramural R & D is found in the top four states. However, the concentration of this R & D in a few places does not necessarily result in measurable regional economic impacts (Holman, 1974; Malecki, 1982). Many of the energy laboratories are located in isolated areas, where little spin-off has occurred. Even in large urban areas, federal facilities do not necessarily lead to further industrial activity. The presence of both Argonne National Laboratory and the Fermilab National Accelerator Laboratory in suburban Chicago has had no discernible effect on the industrial mix of R & D in the Chicago region, which remains heavily tied to food, machinery, and other sectors unrelated to energy (Malecki, 1979).

The spin-off phenomenon and the notion that technological complexes will evolve out of R & D centers pervades the American literature on R & D. At the same time, although Silicon Valley near San Francisco and Route 128 near Boston have served as examples of technology-based regional economic development, there is little evidence of systematic government influence (Bollinger, Hope, and Utterback, 1983).[2] Large government facilities for space and energy R & D were often sited in the South or other areas not known for their research expertise. There are few success stories to recount, and many more examples of areas which failed to generate sustained research complexes. Holman (1974, 270-1) points out that even some centers of large-scale government defense and space R & D, such as Houston, Seattle, and St. Louis, have failed to develop R & D complexes. The principal effects of the space program on its areas of influence tended to be short-term local growth, somewhat higher incomes, and increased dependence on federal spending (Holman, 1974). The space program, like public sector R & D generally, has relied on large firms, which tend to spread very little of their R & D to geographically wider regions (Malecki, 1982, 1984).

As Buswell (1983, 16-17) notes, the circular and cumulative effect associated with the development of R & D complexes is likely to prove extremely variable from place to place. This point has been echoed by Bollinger, Hope, and Utterback (1983). Spin-offs, typically an important part of the development of such complexes, are poorly understood and tend to be more related to the characteristics of products produced in an area than to local environmental conditions, such as government R & D or university research (Garvin, 1983). The Research Triangle area in North Carolina, one of the other 'success stories' among R & D complexes, is a case of government influence in health and medical fields rather than in defense, space, or energy. The location of several major federal laboratories and branches of some large corporations has failed to result in the profusion of spin-offs found in California or Massachusetts (Farrell, 1983).

It is difficult to measure directly the impact of government R & D on regional economic performance (Malecki, 1982), since such R & D is related to other actions by the public sector that affect regional economic growth unevenly. However, the support by government agencies of R & D and the purchase of new products flowing from it has been cited as the cause of the phenomenal growth of new firms in Silicon Valley during the 1960s, to a degree unmatched elsewhere since (Hanson, 1983; Schnee, 1978; Wilson, Ashton, and Egan, 1980). Although military purchases now overwhelmingly favor large firms, there remains a general association between defense spending and growth in high-technology industry (Glasmeier, Hall, and Markusen, 1983). The closest possible explanation for this growth seems at the moment to be one relying on the notion of agglomeration of technical labor necessary for R & D and the subsequent economies available for firms conducting R & D in such concentration (Oakey, 1983).

Firms which perform R & D, either internally or for the government, need to attract and keep a pool of scientists, engineers, and technicians. This labor force is an important asset of such firms and as a location factor contrasts markedly with production labor. The availability of an R & D work-force largely determines location; the relative wage or salary level is unimportant and firms, in their drive to access large pools of research workers, seem to accept willingly the high wage costs. The R & D employees themselves, then, largely determine where R & D can take place, whether public or private in nature. Generally, these are urban areas that have the amenities sought by research personnel, including major universities, good airline connectivity, and urban cultural activities (Browning, 1980; Dorfman, 1983). Places with these characteristics tend to be large cities or places with a long-standing R & D focus or, at least, such a reputation (Oakey, 1983). For the R & D professional, such locations also have the added advantage of multiple employment opportunities in the same urban area, thus allowing job mobility without requiring long-distance moves. Even major government R & D contractors tend to keep their research teams at their more attractive corporate locations, and divert relatively little effort to military installations, although military personnel are stationed at such places (Table 5.5). Law (1983) shows similar patterns in British military R & D.

4. Conclusions

It is difficult to make 'broad-brush' statements about government R & D and its effects on the economic performance of regional economies. The distribution of government R & D in the United States has important impacts on the economies of some regions. However, it has taken a combination of factors, many of them related to the private sector, to generate the high-technology regional economies that are the goal toward which many regions aspire. The evidence reviewed in this chapter suggests that university R & D, although only a small part of the total outlays from government, has a disproportionate effect on the regions in which it is concentrated. Other areas may benefit from certain large-scale government R & D in certain specialized facilities, such as the NASA complex and

Table 5.5 R & D locations by SMSA of major
defense R & D contractors, fiscal year
1980 ($m)

Boeing	($m)
Total R & D	782.5
Huntsville, Alabama	5.3
Santa Barbara, California*	0.0
Wichita, Kansas	314.4
Albuquerque, New Mexico*	0.2
Houston, Texas	0.3
Seattle, Washington	462.2
McDonnell Douglas	
Total R & D	620.7
Los Angeles, California	88.8
St. Louis, Missouri	529.8
Albuquerque, New Mexico*	2.1
Martin Marietta	
Total R & D	518.9
Santa Barbara, California*	54.1
Denver, Colorado	166.4
Orlando, Florida	288.7
Baltimore, Maryland	0.4
Boston, Massachusetts	9.3
Rockwell International	
Total R & D	502.5
San Francisco, California	0.1
Los Angeles, California	396.0
Cedar Rapids, Iowa	39.5
Columbus, Ohio	60.6
Dayton, Ohio*	0.1
Dallas-Fort Worth, Texas	6.2
Hughes Aircraft	
Total R & D	493.3
Tucson, Arizona	16.2
Los Angeles, California	474.9
San Diego, California	0.2
Dallas-Fort Worth, Texas	1.9

* Designates military base location.
Source: US Department of Defense (1981), *500 Contractors Receiving the Largest Dollar Volume of Prime Contract Awards for RDT&E, Fiscal Year 1980;* Washington, DC, US Department of Defense, p. 11.

its spin-offs near Cape Canaveral. These facilities and their impacts, however, are far from ubiquitous, and cannot be relied upon for regional economic vitality.

Government R & D and the procurement of high-technology products have their greatest impact in places where corporate performers agglomerate (such as Los Angeles and San Francisco), or where universities have large-scale, federally funded research programs. In fact, firms seem to be attracted to university research on a regional or state level as well as to specific institutions. These regional impacts are largely implicit or coincidental outcomes of federal R & D programs, and little concern is expressed over their geographic concentration. Explicit locational impacts take place from R & D at federal intramural facilities and FFRDCs. The impact of the former is mostly felt in the Washington, DC, region (including Maryland and Virginia) where over one-third of all intramural R & D is conducted. The impacts of FFRDCs are minimal for those located in small, isolated cities where little other R & D and few other technological determinants are present (e.g. Los Alamos, New Mexico and Idaho Falls, Idaho). The same is largely true of those in relatively small metropolitan areas (e.g. Albuquerque, New Mexico or Knoxville, Tennessee), at least in comparison with major R & D complexes. The regional effects of these government facilities—including the ease of recruiting personnel—are probably less than those of other government laboratory sites in the Boston, Los Angeles, Pittsburgh, and San Francisco urban areas.

Public sector R & D, then, may be both over-concentrated in a few firms and a few locations and over-specialized in defense-related sectors to have widespread economic effects. In regions that are the home of several research universities and attractive residential and recreational environments, the effect of government R & D is perhaps many times that in other regions. Regional economic models are unable to quantify this, mainly because of difficulties in measuring long-term economic potential and its effects. In addition, the bulk of American federal R & D goes to military–space–energy fields which are dominated by large firms that rarely disperse benefits greatly from existing concentrations of facilities and personnel (Malecki, 1982, 1984). The evidence available from the UK suggests similar concentration (Law, 1983).

The role of public sector R & D, while indirect, is of great importance because of government support of university research. Industrial R & D will always depend to a large degree on university research, and a geographically dispersed pattern of high quality institutions in different fields of research suggests a broad infrastructure for R & D. It should be recognized, however, that just as corporate basic research is different from product development, product improvement, and high-volume production, universities vary greatly in the range, quantity, and quality of their research, as well as in their degree of contact with industry. Not all universities attract industrial R & D, and the best schools and programs will continue to attract a more than average share.

Regional and local economic historical evolutions generate a complex set of conditions that together either serve to attract or fail to hold a healthy balance of new and old industries and firms (Hekman and Strong, 1981). It is this complex interaction of many factors that has rendered elusive a full understanding of regional economic development. Public sector R & D is but one of these factors, albeit perhaps the critical one in specific locations.

Notes

1. The effects on regional growth and income of government R & D appear to be minimal, perhaps because of the relatively small proportion of regional income that comes from R & D (Malecki, 1982).
2. The initial impetus for the Boston and Silicon Valley complexes was the military-related research programs begun just after World War II, which were quite concentrated in few universities. These may have provided a set of unique initial advantages that has not been duplicated elsewhere (Dorfman, 1983; Hanson, 1982). It is also possible that no government R & D programs in other regions have approached the intensity or degree of technology advance of that period.

References

Bollinger, L., Hope, K., and Utterback, J. M. (1983). 'A Review of Literature and Hypotheses on New Technology-Based Firms,' *Research Policy*, 12:1, 1–14.

Browning, J. E. (1980). *How To Select a Business Site* (New York, McGraw Hill).

Buswell, R. J. (1983). 'Research and Development and Regional Development: A Review,' in A. Gillespie (ed.), *Technological Change and Regional Development* (London, Pion), pp. 9–20.

Cooper, A. C. (1971). 'Spin-Offs and Technical Entrepreneurship,' *IEEE Transactions on Engineering Management*, 18, 2–6.

Dorfman, N. S. (1983). 'Route 128: The Development of a Regional High Technology Economy,' *Research Policy*, 12:6, 299–316.

Farrell, K. (1983). 'High-Tech Highways,' *Venture*, 5 (September), 38–50.

Fusfeld, H. I., Langlois, R. N., and Nelson, R. R. (1981). *The Changing Tide: Federal Support of Civilian-Sector R & D* (New York, New York University Center for Science and Technology Policy).

Gansler, J. (1980). *The Defense Industry* (Cambridge, MIT Press).

Garvin, D. A. (1983). 'Spin-Offs and the New Firm Formation Process,' *California Management Review,* 25:2, 3–20.

Glasmeier, A. K., Hall, P., and Markusen, A. R. (1983). 'Recent Evidence on High-Technology Industries' Spatial Tendencies: A Preliminary Investigation,' Working Paper 417 (University of California, Berkeley, Institute of Urban and Regional Development).

Glickman, N. J. (ed.) (1980). *The Urban Impacts of Federal Policies* (Baltimore, Johns Hopkins University Press).

Hanson, D. (1982). *The New Alchemists: Silicon Valley and the Microelectronics Revolution* (Boston, Little, Brown).

Hekman, J. S. and Strong, J. S. (1981). 'The Evolution of New England Industry,' *New England Economic Review* (March/April), 35–46.

Holman, M. (1974). *The Political Economy of the Space Program* (Palo Alto, Pacific Books).

Kochanowski, P. and Hertzfeld, H. E. (1981). 'Often Overlooked Factors in Measuring the Rate of Return to Government R & D Expenditures,' *Policy Analysis,* 7, 153–67.

Law, C. M. (1983). 'The Defence Sector in British Regional Development.' *Geoforum,* 14:2, 169–84.

Lee, K. H. (1983). 'Interstate Variations in Manufacturing Growth in the U.S.: Dual Roles of Technological Change,' unpublished Ph.D. dissertation, University of Oklahoma.

Malecki, E. J. (1979). 'Locational Trends in R & D by Large U.S. Corporations, 1965–1979,' *Economic Geography,* 55, 309–23.

Malecki, E. J. (1981). 'Science, Technology, and Regional Economic Development: Review and Prospects,' *Research Policy,* 10, 312–34.

Malecki, E. J. (1982). 'Federal R & D Spending in the United States of America: Some Impacts on Metropolitan Economies,' *Regional Studies* 16, 19–35.

Malecki, E. J. (1984). 'Military Spending and the US Defense Industry: Regional Patterns of Military Contracts and Subcontracts,' *Environment and Planning C: Government and Policy,* 2, 31–44.

Norman, C. (1979). *Knowledge and Power: The Global Research and Development Budget,* Worldwatch Paper 31 (Washington, DC, Worldwatch Institute).

Oakey, R. P. (1983). 'New Technology, Government Policy and Regional Manufacturing Employment,' *Area,* 15:1, 61–5.

OECD (1981). *Science and Technology Policy for the 1980s* (Paris, Organization for Economic Co-operation and Development).

Pavitt, K. (1979). 'Governmental Support for Industrial Innovation: The Western European Experience,' in R. Johnston and P. Gummett (eds), *Directing Technology* (New York, St. Martin's Press), pp. 21–40.

Scherer, F. M. (1982). 'Inter-Industry Technology Flows in the United States,' *Research Policy,* 11, 227–45.

Schnee, J. E. (1978). 'Government Programs and the Growth of High-Technology Industries,' *Research Policy,* 7, 2–24.

Steed, G. P. F. and DeGenova, D. (1983). 'Ottawa's Technology-Oriented Complex,' *Canadian Geographer,* 27:3, 263–78.

Van Duijn, J. J. and Lambooy, J. G. (1982). 'Technological Innovation and Regional Economic Growth: A Meso-Economic Analysis,' Research Memorandum 8207 (University of Amsterdam, Department of Economics).

Wilson, R. W., Ashton, P. K., and Egan, T. P. (1980). *Innovation, Competition, and Government Policy in the Semiconductor Industry* (Lexington, Mass., Lexington Books).

6 The Diffusion of New Production Innovations in British Industry

D. C. GIBBS and A. EDWARDS
Centre for Urban and Regional Development Studies
University of Newcastle upon Tyne

1. Introduction

One of the more enduring features of the British economy over the past half-century has been the poor industrial and economic performance of the peripheral regions. Unemployment remains consistently high in these regions and household incomes are below the national average. British regional policy has sought to ameliorate these disadvantages by influencing the mobility of industry from prosperous to backward regions and improving the performance of indigenous industry in Development Areas through the subsidy of capital investment, particularly the blanket Regional Development Grant (Cameron, 1979; Rees and Miall, 1981). During the present recession, regional inequalities are likely to persist or worsen, since manufacturing output has fallen at a faster rate than that in the service or public sectors, and peripheral regions are especially dependent upon manufacturing jobs. Moreover, in recent years there has been a reduction in the extent of government regional assistance, in terms of expenditure and areal coverage. If the depressed regions are to improve their present position, new approaches to regional problems are required.

This is not, however, to denigrate past policy. Regional Development Grants are estimated, for example, to have substantially raised the level of investment in the Assisted Areas. Moore, Rhodes and Tyler (1977) go so far as to suggest that industrial investment in the Development Areas may be 30 per cent higher than might have been expected without the influence of policy. However, other studies suggest that capital/output ratios and incremental capital/output ratios are higher in the Assisted Areas than elsewhere (Northern Region Strategy Team, 1977; Cameron, 1979). In other words, a higher level and rate of investment is required in Development Areas than elsewhere in Great Britain to raise output by a given level in any time period.

* The research on which this chapter is based was funded jointly by the United Kingdom Department of Trade and Industry and by the Commission of the European Communities. However, the views expressed and conclusions reached are not necessarily those of the funding bodies. The authors would like to thank Neil Alderman for undertaking the logit analysis and Alfred Thwaites for his valuable comments. All errors and omissions remain the responsibility of the authors.

Differential productivity may be one of the keys to the explanation of regional variation in industrial performance. If this is the case, then investment in new technology, rather than the encouragement of investment *per se,* may provide regional policy with a mechanism for encouraging industrial growth in the peripheral regions through improving capital/output ratios and labour productivity. Although work by Swan and Kovacs (1980) argues that regional inequalities must be reduced through productivity improvements, there is no mechanistic way of improving productivity across-the-board in all industries. Indeed, opportunities to improve productivity and competitiveness may be limited by regional variations in the adoption and diffusion of innovations such as new machinery, production techniques, and management systems.

The importance of technological change to the development of national economies and to the survival and improved competitive position of manufacturing enterprises has long been established (Mansfield, 1968; Schmookler, 1972; Freeman, 1974). It is recognized that the strength of many of Britain's major trading competitors lies to a greater or lesser extent in technological capabilities incorporated within new products and production processes; factors which complement other expertise in organization and marketing. If British industry is to arrest decline by improving its export performance and resisting import penetration under current conditions of rapid technological change, it must incorporate advanced technology in its products and processes at least as rapidly as its major competitors. Indeed, Britain has been unfavourably compared to other OECD countries whose 'industries operate on the whole closer than British industry to international best-practice techniques, which use factors of production more efficiently and therefore allow higher wages at a given level of profitability' (Pavitt, 1981, 88).

If this process is essential at a national level then it is equally important at the subnational or regional level within nations. At the regional level, improvements in technological performance can contribute, in the longer term, to self-sustained regional economic growth. In the absence of other strong equalizing forces, unequal technological advance between regions will be to the long-term detriment of those which lag behind as their products become obsolete and processes uncompetitive. The implications of technological change for industry in depressed areas are at least as important as those for industry in prosperous regions. The adoption of new technology in the peripheral regions must occur at a rate equal to the national average if regional disparities are not to increase. Moreover, new technology may offer a way to ameliorate inequalities if depressed regions have adoption rates above those in other regions: a region can only maintain a high level of growth if, in the long run, firms estab-lished within it are able and prepared to adjust continually to technological change (Pederson, 1971).

However, conditions favouring technological change such as research and development (R & D) expenditure and high levels of capital investment have been shown to vary between regions (Buswell and Lewis, 1970; Thwaites, Oakey and Nash, 1981) and it has been hypothesized that such variations may affect technological innovation (Ewers and Wettman, 1980). Few studies of technological innovation have focused upon any specific regional differentials; a notable exception was the early work of Mansfield (see Mansfield *et al.,* 1971) who discovered time lags in the adoption of numerically controlled machine tools between states in the USA. Specific regional differentiations were discovered in a study carried out for the Economic Council of Canada into the rates of adop-tion of various techniques in Canadian industry (Martin *et al.,* 1979). In the British context, research has shown that regional variations in the introduction of varying levels of technical innovation, and especially product innovation, occur to the advantage of the South East and to the detriment of the peripheral areas of the country, notably the Northern region, Scotland and Wales (Thwaites, Oakey and Nash, 1981).

New technology may be either developed within the firm or adopted by it from commercially available sources. Previous research concluded that while product innovations were closely tied to the developing firm to gain maximum commercial advantage, the vast majority of process innovations were readily available *at a price* to those firms who wished to take advantage of them. Thus, 'provided capital is available, new pro-cess machinery can be readily purchased from outside the local area or abroad with little difficulty' (Oakey, Thwaites and Nash, 1982, 1,075). This introduces the important concept that technological change is not a homogenous phenomenon but can be divided, in so far as manufacturing industry is concerned, into a number of elements including varying degrees of product and process change. The stimulation of each element may call for a different approach and also achieve different ends in terms of economic and occupational activity, and achieved in different time periods (Oakey, 1983).

Although the encouragement of product innovation *de novo* is an important long-term strategy, a simpler alternative to improve productivity within the peripheral regions is offered by the diffusion of commercially available process-control mechanisms. The advantage of encouraging the adoption of process-control technologies is that they are widely available 'off-the-shelf' as products promoted by manufacturers, such as computer-controlled machine tools and test and design equipment, as well as computer equipment for more traditional activities such as accounting and stock control. The adoption and diffusion of such proven technologies represents a relatively low-cost, short time span, and lower-risk approach to technological change. These features also provide the opportunity for policymakers to encourage adoption as part of an overall regional policy aimed at improving the productivity and competitiveness of peripheral regions. Such adoption could raise the level of technological awareness and knowledge in the establishment or enterprise which might in turn lead to further adoption of higher technologies and eventually to more substantial innovation. Thus a strategy of encouraging the adoption of high-technology process innovations may have an educational 'spin-off' effect which could subsequently aid the longer-term processes of fundamental technical change. In order to devise policy to achieve this more limited and shorter-term objective, we must possess some understanding of factors which promote or inhibit the regional adoption process. Although much research has sought to explain variations in the rate and level of adoption of manufacturing techniques amongst nations and enterprises (Boretsky, 1975; Mansfield, 1968), little is known of the local or plant level factors essential to understanding at the regional level. This paper attempts to rectify this situation by providing evidence on the adoption and diffusion of a number of high-technology process innovations in the British economy and the use of high-technology sub-components (i.e. microprocessors) in products.

2. Survey methodology

Although previous research by Thwaites, Oakey and Nash (1981) had revealed few regional differences in the adoption of process technologies acquired from sources external to the plants concerned, their work had defined process innovation as any production process new to a particular establishment. Obviously such a definition provides the possibility of encompassing a wide range of production processes incorporating high and low technologies. In an attempt to avoid this problem, it was thought essential in the research discussed below to concentrate upon a number of substantive changes in production technologies, which had a bearing upon the performance of a wide range of industries.

Manufacturing industry in recent years has experienced rapid technological changes which have focused upon improving methods of control and monitoring production processes utilizing the advances in micro-electronics that have facilitated the swift and accurate processing of very large quantities of data. Computers, and more recently microprocessors, have begun to find a wider range of applications within industry, not only enabling management to control and direct production and associated requirements (stocks, orders, etc.), but also to control production machinery directly. This study attempts to trace the interregional diffusion of a number of these developments:

— computerized, numerically-controlled machine tools;
— computer-based production planning and design facilities; *and*
— the utilization of microprocessors both in manufacturing processes, such as in test and safety equipment, and in products.

Although the potential for using computers in such applications has existed in industry for some time, major problems have arisen in the installation of computers and microprocessors into a shop-floor environment because of the need to design suitable sensor mechanisms and to adapt sensitive microelectronics to work in adverse shop-floor conditions (Bessant *et al.,* 1980). The increased use of these new techniques also reflects rising demand for batch production to enable products to be tailored to customer demands, as well as facilitating the ability of manufacturers to respond more quickly to the changing requirements of the market. The trend towards the increased use of automation has been apparent for some time (Bylinsky, 1975); the advantages of these specific process-control technologies is that they can be adopted on an *ad hoc* basis, at the same time allowing for the prospect of linking the computer control systems together to achieve integrated manufacturing systems.

The selection of the techniques outlined above and the industries for study was an interactive process. The techniques were selected on the basis that they introduced fundamental as opposed to minor incremental change, were economically significant and had a comparatively recent diffusion pattern. They were also required to be appropriate to a number of industries which, in order to examine regional variation, were substantially represented in the majority of economic planning regions within Great Britain. Figure 6.1 shows those regions and those areas in receipt of government aid over the period of adoption. In total, nine industrial sectors (defined at the minimum list heading (MLH) level) were chosen. These sectors were:

MLH

331   Agricultural machinery
332   Metalworking machine tools

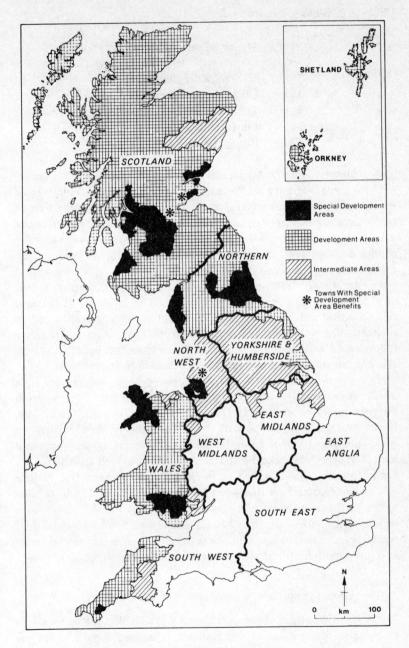

Figure 6.1 British Regional Development Aid (1979)
Source: A. Townsend, Department of Geography, University of Durham.

333	Pumps, valves and compressors
336	Contractors' plant and equipment
337	Mechanical handling equipment
339	Other mechanical engineering
341	Industrial plant and constructional steelwork
361	Electrical machinery
390	Engineers' small tools and gauges

A relatively short postal questionnaire was dispatched to all known establishments operating in the nine sectors to determine adopting and non-adopting establishments for each technique as well as the date of first adoption where appropriate. The questionnaire also sought to obtain a limited amount of information concerning the size and ownership of the establishment, the major products manufactured, market interdependence and the extent of R & D activity.

Of the 4,900 plants approached with a postal questionnaire, a total of 1,234 usable responses were obtained which, given a high level of firm closures and movement out of manufacturing, was estimated to represent 40 per cent of the identified population. A subsequent telephone follow-up of a 10 per cent sample of non-respondents indicated reasonable correspondence with the respondent subset of the population.

The postal questionnaire was also intended to provide a base from which systematically derived samples of establishments for in-depth interview could be selected. The interview survey was designed to explore, in greater depth than was possible in the postal survey, the characteristics of adopting and non-adopting establishments, including their approach to technological change and investment generally, as well as their reasons for adoption or non-adoption of the specified techniques. Due to time and resource constraints, the interviews were limited to 130 in number in four of the ten economic planning regions of Great Britain; the Northern region, the South East, the West Midlands and Scotland. These four regions were chosen to represent a range of diverse economic environments in terms of both historical and current industrial contexts.

3. The regional perspective to the adoption of technology

The analysis from the postal questionnaire reveals that spatial variations in the rate of adoption of the innovations covered by the survey are relatively small. Table 6.1 shows that, in relation to national levels of adoption, the greatest spatial variation is in the incorporation of microprocessors into products, and the least spatial variation in the adoption

Table 6.1 Adoption of new technology by Assisted Area status

	Percentage of respondents in each area having adopted				
	CNC	Computers in commercial use	Computers in manufacturing and design	Micro-processors in manufacturing processors	Micro-processors in products
Development Areas	21.3	60.9	28.2	10.1	13.9
Intermediate Areas	25.4	70.3	31.1	12.6	20.2
Non-Assisted Areas	27.0	64.5	29.8	11.6	22.4
South East	24.5	62.6	22.8	11.4	22.7
Great Britain	24.8	64.3	28.3	11.3	20.1
Spatial variation coefficient*	22.9	14.6	29.3	22.1	43.8
	$N = 1,234$	$N = 1,188$	$N = 1,234$	$N = 1,173$	$N = 1,002$

* Calculated as: $\dfrac{\text{maximum \% − minimum \%}}{\text{mean \%}} \quad 100$

Source: Thwaites, Edwards and Gibbs, 1982.

of computers for commercial purposes. Although the spatial variations are small, they are, however, consistent. The Development Areas record below-national-average adoption rates for all the techniques examined, with the exception of computers for manufacturing and design purposes. Moreover, by comparison with the rest of the country, the Development Areas displayed significantly lower levels of adoption for microprocessors in their products, a factor which may have a depressing effect upon the market competitiveness of the products of such firms and their host regions.

Adoption, however, is a temporal process and these spatial results only indicate the situation extant in 1981. Figure 6.2 provides details of spatial variations in the pattern of diffusion over time for each of the techniques in relation to Assisted Area status. For the adoption of CNC machine tools, Figure 6.2(i) shows that, although by 1981 the Development Areas had a record of adoption lower than the national average, this would appear to be a situation which arose after 1980. Until that time the Development Areas appear to have had their fair share of innovative establishments. In the other areas any deflection from the upward trend in levels of adoption comes later and, by mid-1981, seemed less severe. Whether this indicates that the effective saturation level in the Development Areas will be lower than elsewhere, or that Development Areas are suffering particular hardship in the current recession, leading to a decline in investment, is not at all clear. These changes in trend are not as apparent in the other Non-Assisted Areas outside the South East, where a high level of adoption has been sustained.

There are few spatial variations in the pattern of diffusion over time for computers for commercial purposes (Fig. 6.2(ii)). This was also found to be the case for computers used for design and manufacturing purposes, with the exception of the South East which consistently records a lower-than-national-average level of adoption throughout the 1970s (Fig. 6.2(iii)). The adoption rate for computers in commercial use is much higher in all areas than the rate for computer numerically controlled (CNC) machine tools. At the lowest level, 21.3 per cent of establishments in Development Areas had adopted CNC by 1981, while 61 per cent had adopted computers for commercial use (see Table 6.1). This probably reflects differences in the applicability of different technologies, as well as the stage reached in the diffusion process for individual techniques. The real period of take-off for computers seems to have occurred in the early to middle 1970s for commercial uses, and late 1970s for manufacturing and design use, coinciding with the availability of reasonably cheap and powerful mini- and microcomputers, which put the technology within financial

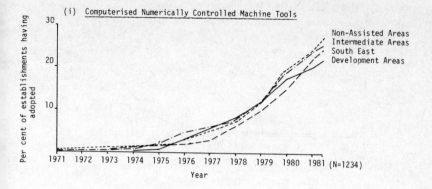

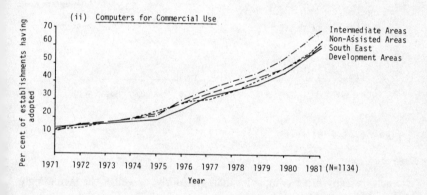

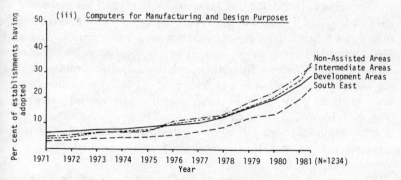

Figure 6.2 The diffusion of new technology by location
Source: Thwaites, Edwards and Gibbs, 1982.

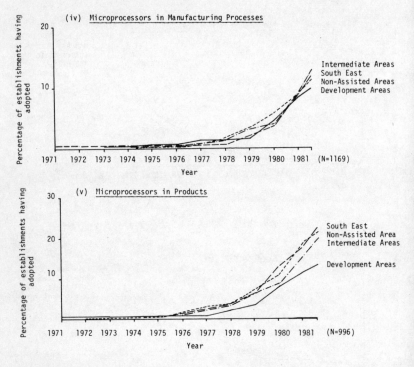

Figure 6.2 (*continued*)

reach of many more companies than previously, as well as the increasingly 'user friendly' nature of such systems.

Perhaps because of the relatively low level of adoption of microprocessors in processes (excluding CNC) reached by 1981 (11.3 per cent nationally), little spatial variation in the levels of adoption over time is observed (Fig. 6.2(iv)). Since 1980, however, there is some indication that, as with CNC, the rate and level of adoption in the Development Areas has decreased relative to other locations. By contrast, industry in the Development Areas has adopted microprocessors into their products at a consistently lower level at each point in time than establishments located elsewhere (Fig. 6.2(v)). This is particularly in contrast to the Non-Assisted Areas and the South East region. If anything, the situation has further deteriorated since 1980 with Development Area adoption taking place at a much slower place than elsewhere.

At a more spatially disaggregated level, Table 6.2 shows an index of adoption for the standard economic regions. The table shows that certain

Table 6.2 Adoption of new technology by economic region (Index of regional adoptive performance, GB = 100)

Economic planning region	CNC machine tools	Computers for commercial activities	Computers for manufacturing and design purposes	Micro-processors in manufacturing processes	Micro-processors in manufactured products
East Midlands	112.9	110.1	108.5	112.4	117.9
West Midlands	104.4	106.5	125.1	105.3	127.4
South West	119.0	90.2	91.9	96.5	89.1
East Anglia	96.4	97.8	92.2	99.1	113.4
Yorkshire and Humberside	113.3	116.6	117.7	122.1	99.5
South East	98.8	97.4	80.6	100.9	112.9
North West	91.9	101.7	102.1	100.0	102.0
Scotland	89.1	99.7	91.9	112.4	94.5
Northern	83.1	97.2	101.1	83.2	60.7
Wales	83.9	74.4	117.7	38.9	25.4

Source: Gibbs, Edwards and Thwaites, 1982.

regions, in particular Yorkshire and Humberside and the East Midlands, performed consistently better than others. Peripheral areas, especially the Northern region and Wales, show a poor performance in adopting the techniques, notably in the incorporation of microprocessors into products.

The South West has the highest proportion of establishments which have adopted CNC, this possibly being related to the presence of the aerospace industry around Bristol, an important early user of the technique, providing an example for potential users and a network of suppliers. Yorkshire and Humberside and the West Midlands possess the best record of establishments adopting computers and microprocessors. The only peripheral region which features in the top three of these technological developments is Scotland, which ranks highly in the use of microprocessors. This may reflect the presence in Scotland of American and Japanese electronic firms, such as Motorola, IBM, Hewlett-Packard, Nippon Electronics, and National Semiconductor. Such firms have located mainly in the Scottish new towns, particularly Livingston, East Kilbride, and Glenrothes, encouraged by financial and other incentives available through central and local government:

> In the past 18 months more than £180m has been invested in Scottish microelectronics by overseas firms. There are now 38,000 people working in electronics in Scotland and the industry has £850m annual sales (*The Sunday Times*, 1981.)

Scotland is the only development area which ranks highly in any of the techniques under study and displays signs of a transformation from older to newer high-technology industries which accords with other studies into high-technology industry (Oakey, 1984). Even so, Scotland's overall record of adoption falls well below those levels achieved in regions in the Non-Assisted and Intermediate Areas.

4. Factors affecting adoptive behaviour

Previous research has argued that the conditions favouring technical change may vary systematically between regions (Thwaites, 1978; Goddard *et al.*, 1979; Ewers and Wettmann, 1980). The suggestion is that these variations arise not only as a result of the location of the region but also because of the nature of the enterprise (or parts of enterprises) operating in the area; characteristics which in combination influence the local climate for innovation. It was thought important, therefore, to identify the characteristics of manufacturing establishments, including regionally differentiated aspects of those characteristics, which influence the speed and level of take-up of the process technologies under investigation.

Perhaps not surprisingly the level of adoption in an establishment was found to be influenced by the industry for which it produces. The industries of metalworking machine tools (MLH 332) and pumps, valves and compressors (MLH 333) displayed a high level of adoption for all the techniques examined (Table 6.3). At the other end of the scale, these same techniques had made relatively little impact by 1981 on agricultural machinery (MLH 331), engineers' small tools and gauges (MLH 390) and industrial plant and constructional steelwork (MLH 341). There were obviously also variations amongst industries in the level of adoption of *individual* techniques rather than of the techniques in aggregate. This point is perhaps best illustrated in contractors' plant and equipment (MLH 336), which has a high level of adoption of computers for design and manufacturing purposes but at the same time a relatively low level of adoption of microprocessors. Such inter-industry differences also formed the basis of plant-level differentiation in the interview survey. For example, a number of nonadopters had a substantial proportion of their output in very large batch sizes and/or low tolerance levels which had discouraged the introduction of CNC machine tools, which are best suited to small-batch, high-quality production.

The industrial structure of regional industry was therefore expected to influence the level of adoption of technology in the area. Further analysis of this, and other characteristics, was undertaken in two main ways. Firstly, a procedure was adopted using standardized share analysis (Malizia, 1978) to examine the observed variations in the adoption of new technology. This method divides regional variations into two component parts; those resulting from spatial variations in *structural conditions,* such as the industrial composition of an area, the size distribution of its establishments etc.; and secondly, those resulting from other factors affecting the *performance* of individual establishments in that region and within that structure. The data were analysed by logit analysis, a multivariate technique which attempts to overcome the difficulties inherent in bivariate analysis and, by multivariate techniques, attempts to approach the rigour of multiple regression models using categorical data with a dichotomous variable (Wrigley, 1980; Upton, 1981; Alderman *et al.,* 1982).

While Development Areas were discovered to possess fewer establishments in highly adoptive industries and more establishments in less adoptive industries, industrial structure was not a significant determinant of

Table 6.3 Adoption of new technology by industrial sectors

MLH	Description	Percentage of respondents					
		CNC	Computers in commercial use	Computers in manufacturing and design	Computers in manufacturing processes	Microprocessors in manufacturing processes	Microprocessors in products
331	Agricultural machinery	10.0	68.4	15.0		5.4	12.5
332	Metalworking machine tools	39.6	74.5	34.2		15.7	46.5
333	Pumps, valves and compressors	43.0	77.6	36.4		17.1	15.7
336	Contractors' plant and equipment	27.3	71.4	40.9		4.5	21.1
337	Mechanical handling equipment	15.4	72.0	32.1		11.6	35.7
339	General mechanical engineering	26.4	64.3	26.4		12.4	25.3
341	Industrial plant and con- structional steelwork	11.4	58.4	28.2		7.1	10.4
361	Electrical machinery	20.3	70.1	37.4		15.3	18.6
390	Engineers' small tools and gauges	22.9	14.5	14.6		6.3	11.3
—	Other mechanical engineering industries	28.2	60.3	23.9		10.5	14.0
	Total: Great Britain	24.8	64.3	28.3		11.3	20.1
		N = 1,234	N = 1,188	N = 1,234	N = 1,234	N = 1,173	N = 1,002

Source: Thwaites, Edwards and Gibbs, 1982.

regional variation. It was not a statistically significant factor in the logit analysis and 'performance' factors were predominant in the standardized share analysis.

The logit analysis suggested that a number of other factors were more strongly associated with regional variations in adoptive behaviour. These were corporate status, establishment employment size and access to research and development facilities. While these factors were found to be important in accounting for the relatively poor performance of the Development Areas, the standardized share analysis also revealed the important of 'performance' factors in exacerbating these problems. In the account of these factors below, evidence is also provided from the interview survey on the differences between adopters and non-adopters.

5. Adoption and establishment characteristics

It proved possible to identify four types of establishment with regard to corporate status; single-plant independent enterprises and those forming part of a multi-plant enterprise, subdivided into group headquarters, regional/divisional headquarters and branch plants. It was hypothesized that those establishments possessing access to the facilities of finance, specialized labour, research and expertise available through a multi-plant enterprise would display higher levels of adoption than single-plant independent enterprises which may be more resource-constrained. Additionally, based upon Vernon's work (1966, 1971), which suggests a hierarchical diffusion process within multi-plant and multinational corporations, it was thought that those establishments with higher status within an enterprise would exhibit the highest levels of adoption.

These hypotheses are generally borne out by the evidence (Table 6.4). While the lowest-status plants of multi-plant enterprises less frequently adopted the techniques under study, group, regional or divisional head-quarters plants proved to be more highly adoptive. Overall, however, single-plant enterprises were the *least* adoptive of all the establishments. Their rate of adoption was consistently below that displayed by the lowest status multi-plants, and those with the least supervisory respons-ibilities in the group. Analysis over time revealed that single-plant enter-prises have consistently adopted all the techniques at a lower level, particularly microprocessors (Fig. 6.3).

Further analysis revealed that company structure in the Development Areas in the sample population was consistently negative (e.g. they possess fewer group plants of higher status and a greater-than-average number of group plants of low status or of single-plant independent enterprises).

Table 6.4 Adoption of new technology by company structure, 1981

	Group headquarters	Regional product headquarters	Other multi-plants	Single plants	Total GB	
Computerized numerical control	31.2	35.3	26.4	18.5	24.8	$N = 1,204$
Computers in commercial functions	81.0	84.9	76.5	46.1	64.3	$N = 1,159$
Computers in manufacturing and design	41.6	39.1	37.4	16.6	28.3	$N = 1,204$
Microprocessors in production processes	20.0	15.6	15.3	6.3	11.3	$N = 1,146$
Microprocessors in products	29.9	27.8	21.1	13.5	20.1	$N = 980$

Percentage of establishments in each cell adopting the technique

Source: Thwaites, Edwards and Gibbs, 1982.

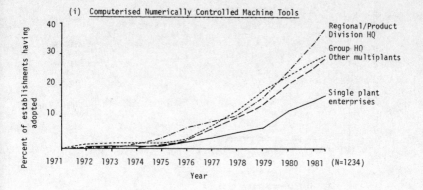

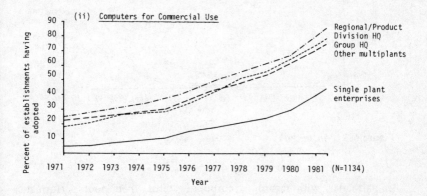

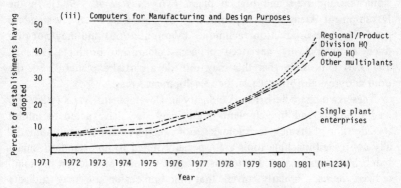

Figure 6.3 The diffusion of new technology by coporate status
Source: Thwaites, Edwards and Gibbs, 1982

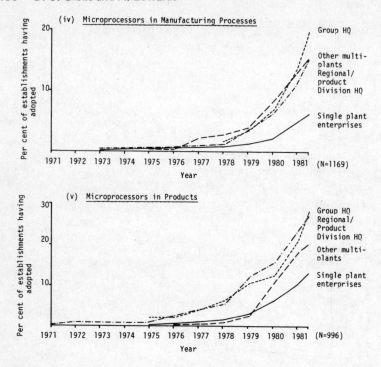

Figure 6.3 (*continued*)

In particular, with regard to corporate status, their poor performance was a consequence of low adoption rates in both regional/product divisional headquarters and branch plants (Thwaites *et al.*, 1982). In the Development Areas, branch plants may be producing relatively standard goods by mass-production techniques (Vernon, 1966) and may not produce technologically advanced or non-standardized products (Keeble, 1972). It is possible that this may provide a partial explanation for the poor adoptive performance of the Development Areas.

These two characteristics of industry in Development Areas (standardized products and branch-plant dominance) may have acted to inhibit the adoption of the new technologies. For example, computerized numerically controlled machine tools are especially suited to non-standard, small-batch metalworking production as opposed to mass production of standardized items. Similarly, given that microprocessor use may indicate technically advanced products or production processes necessitating accurate measurement or quality control, adoption may be less appropriate in situations of standardized production. These product-related factors are interconnected with systems of corporate organization. For a multi-plant enterprise, satisfactory control of each regional unit must be possible and the unit economically viable at a distance from the present plant. Given the concentration of headquarters of multi-plant firms outside the Assisted Areas (Goddard and Smith, 1978), functions which require day-to-day supervision from head office are unlikely to be decentralized to branch plants. In consequence, sites distant from the corporate centre are more likely to be engaged in routine production. In summary, plants in peripheral regions may not engage in production distributed throughout the full product cycle, receiving only those standardized 'end-of-cycle' elements located there by corporate decision-makers. Such decisions determine not only the functions the plants perform, but also the type of investment made, which has obvious consequences for the form of process technology within establishments and regions.

It is thus important to recognize that 'technological diffusion rates in industry are determined in . . . essentially private economies by managerial decisions at the level of individual firms' (Gold, 1981, 254). The problem for Development Areas is that personnel within establishments must have not only the willingness to bring about change, but also the *power* to do so (Townroe, 1975). The interview questionnaire was particularly concerned, therefore, to discover whether levels of autonomy were lower in multi-plant establishments in the two Development Areas studied. Table 6.5 reveals that establishments located in the Northern region and Scotland had considerably less autonomy over matters such as capital expenditure, budgeting and equipment choice than those located in the South East and West Midlands. To some extent, therefore, low adoption levels in the Development Areas may be a partial consequence of corporate structures and power.

Closely related to organizational status is the notion of enterprise size which has been associated with early adoption of new techniques in a number of empirical studies (e.g. Mansfield *et al.*, 1971). This is largely based on the rationale that larger enterprises can more easily accept the risks associated with new techniques, and, through the subdivision of the manufacturing process, reap economies of scale by the use of specialized equipment. The size of resources available to the establishment may also be a crucial factor in the decision to adopt new production processes, as large plants may more readily overcome production barriers or thresholds to adoption which inhibit adoption in smaller establishments.

Table 6.6 shows adoption rates by establishment employment size. Although employment figures are not perfect surrogates for size measured

Table 6.5 Proportion of all adopting multi-plant establishments exercising total control over decision-making*

Areas of control	Northern	Scotland	West Midlands	South East	Total
Production planning	71.4	100.0	100.0	95.0	94.0
Choice of production techniques	57.1	77.8	85.7	90.0	82.0
New customer accounts	27.0	55.5	92.8	80.0	73.5
Equipment choice	42.9	55.5	78.6	85.0	72.0
Wage bargaining	28.6	33.3	78.6	80.0	64.0
Product development	42.8	44.4	80.0	37.0	64.0
Machine replacement	28.6	66.7	64.3	70.0	62.0
Pricing policy	14.3	44.4	85.7	65.0	60.0
Capital expenditure	14.3	0.0	42.8	45.0	32.0
Budgeting	14.3	11.1	28.6	40.0	28.0

* $N = 61$
Source: Thwaites, Edwards and Gibbs.

Table 6.6 Adoption of new technology by establishment employment size

	Percentage of establishments in each cell adopting the technique						
	Number of employees						
	1–99	100–199	200–499	500–999	1,000+	Total GB	
Computerized numerical control	10.9	27.1	41.8	53.4	79.7	24.8	N = 1,207
Computers in commercial functions	42.8	84.9	94.6	97.7	96.8	64.3	N = 1,171
Computers in manufacturing and design	12.3	32.1	44.7	65.9	89.8	28.3	N = 1,213
Microprocessors in production processes	5.0	12.7	19.3	26.2	33.9	11.3	N = 1,148
Microprocessors in products	11.9	24.7	28.8	34.2	47.1	20.1	N = 983

Source: Thwaites, Edwards and Gibbs, 1982.

in terms of value of output (and mask factors such as capital/labour ratios which may vary widely between firms affecting technological choice), it was felt inappropriate to request financial data on a postal questionnaire where experience showed that this would adversely affect the response rate obtained. It is apparent that small establishments with less than 100 employees exhibit a relatively low level of adoption of the techniques under study (Table 6.6). There is also evidence of a growing incidence of adoption as establishment size, measured in employment terms, increases. The difference between large and small plants is perhaps best illustrated in the levels of adoption of CNC. Whereas nearly 80 per cent of establishments employing over 1,000 workers on-site had adopted CNC by the time of the survey, only 11 per cent of those plants employing under 100 workers had done so.

The importance of size as a surrogate for access to resources was made evident from the interview survey in two respects. Firstly, the survey discovered that larger establishments were more likely to obtain investment finance through retained profits or group finance than small establishments which relied particularly upon bank finance. Second, the survey discovered that the ability to systematically monitor and evaluate new technological developments was positively associated with adoption. Small establishments and all non-adopting plants were found to be much less likely to carry out such a systematic monitoring process and, due to resource constraints, it was likely to be less substantial when it was undertaken.

Obviously, the varying establishment size structures of regional industry may therefore influence the aggregate regional rate and level of adoption of these technologies. Analysis of regional variations in size structure revealed that, within the sample population, the Development Areas have a higher-than-average proportion of small establishments and the South East a below-average proportion of the largest establishments. Use of the standardized share analysis suggested that Yorkshire and Humberside, the North West and the East and West Midlands have the advantage of an establishment size structure favourable to adoption, while the Development Areas tend to have adverse establishment size structures. This was frequently compounded by a larger negative performance or regional effect.

The incorporation of relatively high technology into existing production and product frameworks may require considerable research and development effort before introduction can occur, given that it has been suggested that successful innovating firms tend to possess strong 'in-house' professional R & D and have close connections with the outside technological community (Freeman, 1974). Moreover, modifications to innovations may be essential before they are successfully introduced into the workplace. The postal survey revealed a strong relationship between the possession of R & D activities on-site and the adoption of individual techniques (Table 6.7). Moreover, increasing numbers of R & D staff were associated with high levels of adoption, although high proportions of R & D staff to total employment were also associated with increased adoptive performance for microprocessors in products. This contradicts previous research which suggested that process adoption is not heavily reliant on the possession of R & D activity (Thwaites, Oakey and Nash, 1981). However, this difference may be explained by this chapter's concentration upon high-technology processes.

The Development Areas were discovered to be under-represented in terms of the possession of R & D facilities. None of these regions had above 50 per cent of establishments with R & D facilities on-site, compared to an average for other areas of over 70 per cent of establishments with R & D facilities. The standardized share and logit analyses revealed that there was a strong negative structural effect in the Development Areas with regard to R & D facilities.

The results above indicate that establishment characteristics such as industrial sector, employment structure and R & D facilities all affect the propensity and ability to adopt new technology. An examination of the regional variations of these key factors showed that there are a number of structural weaknesses prevalent in the Development Areas which may contribute to their relatively poor performance. Such areas tend to have lower proportions of 'high-technology' industries, larger proportions of branch plants and single-plant enterprises, larger proportions of smaller plants and a lower representation of R & D facilities on-site. In addition, Development Areas have predominantly negative residual characteristics for the adoption of new technologies attributable to 'other regional factors' which tend to aggravate the observed structural weaknesses.

6. The regional factor

These 'other regional factors' may be particularly important at the local level. Establishments in regions with spatially abundant local contact networks may learn of, and be convinced earlier of, the relevance of a particular technique, compared to establishments in more impoverished regions which obtain the technical information from a narrower base and at a later date. In general, the ability of firms to access and use appropriate

Table 6.7 The adoption of new technology by research and development on-site

	Percentage of establishments in each cell adopting the technique			
	With R & D on-site	With no R & D on-site	Total GB	
Computerized numerical control	30.4	14.4	24.8	$N = 1{,}230$
Computers in commercial functions	72.5	49.3	64.3	$N = 1{,}184$
Computers in manufacturing and design	33.5	18.6	28.3	$N = 1{,}230$
Microprocessors in production processes	14.6	5.3	11.3	$N = 1{,}170$
Microprocessors in products	27.9	5.7	20.1	$N = 1{,}000$

Source: Thwaites, Edwards and Gibbs, 1982.

technologies will depend on the degree to which available information is channelled to its decision-makers. 'The diffusion of information therefore becomes a key variable in the diffusion of new techniques' (Thwaites, 1978, 452). However, the strength of information flows may vary over space and information networks may be 'stronger' in certain areas as a consequence of local demand (Pred, 1966; Feller, 1971; Goddard and Morris, 1976).

These factors all form part of the industrial 'milieu' of the establishment whereby some regions may possess more extensive facilities (in the form of research establishments, consultancies, information centres, etc.) than others. The cost of, and access to, technical information may vary considerably between different regions in Britain. Consequently, this may influence the diffusion of innovations by stimulating or inhibiting a firm's adoptive ability.

The interview survey discovered considerable differences between adopters and non-adopters of high-technology production processes in the proportion of establishments which systematically monitored new technological developments of interest to the firm. However, of greater importance to the current argument is that lower proportions of establishments performed such monitoring in Scotland and the Northern region. While both sets of establishments relied upon similar sources for the monitoring process (chiefly technical journals and other relevant publications), adopting establishments had much more diverse means of obtaining information, including an internal R & D commitment, attendance at exhibitions and contacts with customers and consultants. Technology-monitoring policies may also be interrelated with corporate status, given that some branch plant establishments in the Northern region and Scotland were found to rely on intra-corporate technology sources.

That considerable differences may exist in the quality of information networks is also suggested by different perceptions of establishments to the technologies under study. In particular, non-adopters of CNC machine tools and microprocessors in processes commonly underrated the advantages of adoption when compared to the actual experiences of adopters (see Table 6.8). Attitudinal or knowledge problems may therefore play an important role to encourage or discourage adoption. In turn, the phenomena are possibly related to the quality of information sources within a local area. However, it should not be concluded from this evidence that attitudes and awareness are major components of all technological non-adoption. The availability and quality of information may vary widely for specific techniques. While non-adopters of CNC machine

Table 6.8 Comparison of the effect of introducing CNC upon adopters with the perceptions of non-adopters regarding possible adoption*

	Proportion of respondents stating the use of CNC had involved changes or would involve changes					
	Decrease		Remain stable		Increase	
	Adopters	Non-adopters	Adopters	Non-adopters	Adopters	Non-adopters
	%	%	%	%	%	%
Overtime	22.6	35.3	66.1	58.8	11.3	5.8
Jobtime	90.5	30.3	7.9	63.6	1.6	6.0
Setting-up time	66.7	51.7	15.9	41.4	17.5	6.9
Number of jigs/fixtures	74.6	45.0	20.6	50.0	4.8	5.0
Scrap	69.4	36.4	27.4	59.0	3.2	4.5
Stockholding	45.9	34.6	54.1	61.5	0	3.8
Space	38.1	28.0	47.6	66.0	14.3	4.0
Material handling	25.8	40.0	72.6	55.0	1.6	5.0
Inspection	66.7	34.6	28.6	61.5	4.8	3.8

* Adopters: $N = 63$; non-adopters: $N = 49$.
Source: Thwaites, Edwards and Gibbs, 1982.

tools and microprocessors in processes appeared to be relatively unaware of the benefits of these techniques, non-adopters of computer-aided design and microprocessors in products *were* aware of the benefits involved. Non-adoption in these instances were predominantly a result of financial barriers to investment and the technical barriers required to assimilate such technology into production.

Finally, a factor exhibiting considerable regional difference was the part played by available government assistance. With reference to Scotland and the Northern region, the interview survey discovered that extensive use had been made of Regional Development Grants, by both adopters (91.0 per cent) and non-adopters (75.0 per cent) of the technologies under study. Of a total of thirty-five adopters of new technology in these two regions; thirty-two had received RDG—sixteen specifically related to the introduction of CNC machine tools; fifteen for computer hardware for design and/or manufacturing purposes; and only five for the introduction of microprocessor-based equipment (multiple response possible). Although a large majority of firms (68.6 per cent) stated that new technology would have been introduced without the availability of aid, about 50 per cent of respondents said that such aid had brought forward the date of technical change. In the absence of investment incentives it is likely that regional disparities would have been more pronounced (Gibbs, 1984).

In those cases where RDG had not been specifically related to new technological investment, investment incentives were often seen as part of the 'normal' process of investment decision-making. Although technological upgrading was not an explicit aim within the firm, it had become implicit within the investment process. In some instances, although RDG had not brought forward the date of adoption, it had affected the quality (new as opposed to second-hand equipment) and quantity (a greater number of items purchased) of adoptive behaviour. Moreover, it also allowed firms to purchase more expensive and perhaps more sophisticated equipment than would have been possible in the absence of investment aid. Investment incentives in the form of RDG also appeared to have had a beneficial influence upon those establishments which formed part of a multi-plant group. The availability of RDG in a local Assisted Area frequently helped management within a branch establishment to present a better financial case to centralized headquarters for new technological investment at that particular site. Not surprisingly, such RDG availability also led to investment within groups being directed towards particular plants which were eligible for assistance in preference to other non-assisted locations.

7. Summary and conclusion

This paper has demonstrated that, for a number of high-technology-based production processes there exist regional variations in absolute and temporal adoption levels. In particular, this was observed to be detrimental to the British Development Areas, and advantageous to a number of more prosperous regions. To some extent regional differences within the sample population can be explained by the varying structural characteristics of different regions for those factors which influence the adoption process. The Development Areas have less R & D activity, greater proportions of small plants, single-plant independent companies and branch plants, all of which depress their innovation levels. Although the Development Areas also possess an industrial structure which is not conducive to adoption, analysis revealed that this did not significantly explain the poor performance of these regions. This same analysis, however, pointed to a number of 'regional performance factors' which further inhibit technological change. For example, the interview survey provided evidence of a number of issues related to the 'industrial milieu', particularly those factors related to the availability and quality of information regarding new technology, spatial differences in the availability of aid and varying levels of corporate autonomy for establishments belonging to a multi-plant enterprise.

These results would therefore suggest that while Oakey *et al.* (1982) did not detect regional variations in the adoption of externally purchased process innovations in their study, differences do exist in this study between regions in the adoption of certain high-technology process techniques. This has important implications for regional development as 'if new techniques continue to be produced and one region lags behind others in its rate of acceptance of innovations then its economic performance may be inhibited and its pace of economic growth may well be slower than that of its neighbours' (Thwaites, 1978, 451). In policy terms, therefore, this would suggest measures to encourage the adoption of existing techniques in order to improve indigenous competitiveness (Cooke *et al.,* 1984). Such policies would aim to complement longer-term aims of encouraging innovation and research activity and, by faster imitation, to blunt competitive disadvantage (Globerman, 1975). This paper would suggest that a number of infrastructural deficiencies exist in the Development Areas which hinder technological change, including poorer information networks and an inferior level of research activity. Policy could aim to compensate for these and other structural shortcomings through the provision of compensatory mechanisms (Rothwell, 1982).

It is important to recognize, however, that the creation of appropriate infrastructure may not be sufficient in itself. Attention also needs to be devoted not only to the necessary, but also to the 'sufficient conditions' of profitability, market potential and adequate investment finance (Sayer, 1983). While an innovation-orientated regional policy may form an important component of potential indigenous regional economic advance, any such attempts must be clearly rooted in the broader economic context of the national and international economy.

References

Alderman, N., Goddard, J. B. Thwaites, A. T., and Nash, P. A. (1982). 'Regional and Urban Perspectives on Industrial Innovation: Applications of Logit and Cluster Analysis to Industrial Survey Data', Discussion Paper No. 42, Centre for Urban and Regional Development Studies, University of Newcastle upon Tyne.

Bessant, J., Braun, E., and Moseley, R. (1980). 'Microelectronics in manufacturing industry: the rate of diffusion', in T. Forester (ed.), *The Microelectronics Revolution* (Oxford, Basil Blackwell), pp. 198–218.

Boretsky, M. (1975). 'Trends in U. S. Technology: a political economist's view', *American Scientist*, **63**, 70–82.

Buswell, R. J. and Lewis, E. W. (1970). 'The geographic distribution of industrial research activity in the UK', *Regional Studies*, **4**, 297–306.

Bylinsky, G. (1975). 'Here comes the second computer revolution', *Fortune*, **95**, 134–9.

Cameron, G. G. (1979). 'The National Industrial Strategy and Regional Policy', in D. Maclennan and J. B. Parr (eds), *Regional Policy* (Oxford, Martin Robertson).

Cooke, P., Morgan, K., and Jackson, D. (1984). 'New Technology and Regional Development in Austerity Britain: The Case of the Semi-Conductor Industry', *Regional Studies*, **18**, 277–89.

Ewers, H. J. and Wettman, R. W. (1980). 'Innovation Oriented Regional Policy', *Regional Studies*, **14**, 161–79.

Feller, I. (1971). 'The urban location of U.S. inventions 1860–1910', *Explorations in Economic History*, **8**, 285–303.

Freeman, C. (1974). *The Economics of Industrial Innovation* (Harmondsworth, Penguin).

Gibbs, D. C. (1984). 'The Influence of Investment Incentives on Technological Change', *Area*, **16**, 115–20.

Gibbs, D. C., Edwards, A., and Thwaites, A. T. (1982). 'The Diffusion of New Technology and the Northern Region', *Northern Economic Review*, No. 5, 22–7.

Globerman, S. (1975). 'Technological Diffusion in the Canadian Tool and Die Industry', *Review of Economics and Statistics*, **57**(4), 428–34.

Goddard, J. B. and Morris, D. (1976). 'The communication factor in office decentralisation', *Prog. Plann.*, **6**, 1–80.

Goddard, J. B. and Smith, I. J. (1978). 'Changes in corporate control in the British urban system, 1972-1977', *Environment and Planning A*, **10**, 1,073-84.

Goddard, J. B. *et al.* (1979). *The Moblization of Indigenous Potential in the United Kingdom*, Report to the Regional Policy Directorate of the European Community (Centre for Urban and Regional Development Studies, University of Newcastle upon Tyne).

Gold, B. (1981). 'Technological Diffusion in Industry—Research Needs and Shortcomings', *Journal of Industrial Economics*, **29**(3), 247-69.

Keeble, D. E. (1972). 'Industrial movement and regional development in the United Kingdom', *Town Planning Review*, **43**, 3-25.

Malizia, E. (1978). 'Standardised Share Analysis', *Journal of Regional Science*, **18**, 2, 283-91.

Mansfield, E. (1968). *The Economics of Technical Change* (London, Longmans).

Mansfield, E., Rapoport, J., Schnee, J., Wagner, S., and Hamburger, M. (1971). *Research and Innovation in the Modern Corporation* (London, Macmillan).

Martin, F., Swan, N., Banks, I., Barker, G., and Beaudry, R. (1979). *The Inter-regional Diffusion of Innovations in Canada*, Economic Council of Canada.

Moore, B., Rhodes, J., and Tyler, P. (1977). 'The Impact of Regional Policy in the 1970s', *CES Review*, No. 1, CES, London.

Northern Regional Strategy Team (NRST) (1977). *Strategic Plan for the Northern Region*, Vols 1-5, HMSO.

Oakey, R. P. (1983). 'High-Technology Industry, Industrial Location and Regional Development: The British Case', in F. E. I. Hamilton and G. J. R. Linge (eds), *Spatial Analysis, Industry and the Industrial Environment, Vol. 3—Regional Economies and Industrial Systems* (Chichester, John Wiley), pp. 279-95.

Oakey, R. P. (1984). 'Innovation and Regional Growth in Small High Technology Firms: Evidence from Britain and the USA', *Regional Studies*, **18**, 237-51.

Oakey, R. P., Thwaites, A. T., and Nash, P. A. (1982). 'Technological change and regional development: some evidence on regional variations in product and process innovation', *Environment and Planning A*, **14**, 1073-86.

Pavitt, K. (1981). 'Technology in British Industry: a Suitable Case for Improvement', in C. Carter (ed.), *Industrial Policy and Innovation* (London, Heinemann), pp. 88-115.

Pederson, P. O. (1971). 'Innovation diffusion in urban systems', in T. Hagerstrand and A. R. Kulinski (eds), 'Information Systems for Regional Development—A Seminar', *Lund Studies in Geography*, Series B, No. 37, Geerup, Lund, 137-47.

Pred, A. R. (1966). *The Spatial Dynamics of US Urban Industrial Growth, 1800-1914* (Cambridge, MIT Press).

Rees, R. D. and Miall, R. H. C. (1981). 'The effect of regional policy on manufacturing investment and capital stock within the UK between 1959 and 1978', *Regional Studies*, **15**(6), 413-24.

Rothwell, R. (1982). 'The Role of Technology in Industrial Change: Implications for Regional Policy', *Regional Studies*, **16**, 361–9.

Sayer, A. (1983). 'Theoretical Problems in the Analysis of Technological Change and Regional Development', in F. E. I. Hamilton and G. J. R. Linge (eds), *Spatial Analysis, Industry and the Industrial Environment, Vol. 3 — Regional Economies and Industrial Systems* (Chichester, John Wiley), pp. 59–73.

Schmookler, J. (1972). 'Invention and Economic Growth', in Z. Griliches and L. Hurwicz (eds) *Patents, Invention and Economic Change* (Cambridge, Mass., Harvard University Press).

The Sunday Times (1981). 'Moderation pays off for Scotland', 16 August, p. 48.

Swan, N. M. and Kovacs, P. J. (1980). 'Empirical Testing on Newfoundland Data of a Theory of Regional Disparities', Economic Council of Canada, Ottawa.

Thwaites, A. T. (1978). 'Technological change, mobile plants and regional development', *Regional Studies*, **12**, 445–61.

Thwaites, A. T., Edwards, A., and Gibbs, D. (1982). 'Inter-regional Diffusion of Production Innovations in Great Britain', Final Report to the Department of Industry and the EEC, Centre for Urban and Regional Development Studies, University of Newcastle upon Tyne.

Thwaites, A. T., Oakey, R. P., and Nash, P. A. (1981). 'Industrial Innovation and Regional Development', Final Report to the Department of the Environment, Centre for Urban and Regional Development Studies, University of Newcastle upon Tyne.

Townroe, P. M. (1975). 'Branch plants and regional development', *Town Planning Review*, **46**(1), 47–61.

Upton, G. J. G. (1981). 'Log-Linear models, screening and regional industrial surveys', *Regional Studies*, **15**, 1, 33–45.

Vernon, R. (1966). 'International investment and international trade in the product cycle', *Quarterly Journal of Economics*, **80**, 190–207.

Vernon, R. (1971). *Sovereignty at Bay* (Harmondsworth, Penguin).

Wrigley, N. (1980). 'Categorical data, repeated—measurement research designs, and regional industrial surveys', *Regional Studies*, **14**, 6, 455–71.

7 New Technology in the United States' Machinery Industry: Trends and Implications

J. REES
Department of Geography, University of Syracuse,
New York State
and
R. BRIGGS AND D. HICKS
Center for Policy Studies, University of Texas, Dallas

1. Introduction

The industrial restructuring taking place in the United States today has recently been the object of widespread and conflicting policy prescriptions. Recurring recessions through the 1970s and early 1980s have heightened the visibility of the decades-long 'shift to services' and the recent surge of employment growth in high-technology industries. This has produced the curious effect of encouraging many to underestimate the resilience and potential retained by older basic industries. As a consequence, some economic experts advocate the adoption of more explicit industrial, trade and technology transfer policies in order to ensure that this economic restructuring unfolds in an orderly fashion, while others advocate a role for public policy that clears the way for market-orientated adjustments. Those concerned with the social welfare and community dimensions of this transition likewise array themselves along a similar policy continuum.

While single-factor theories of industrial change are properly suspect, a great deal of attention has been accorded technological change in its role as a catalyst of industrial rejuvenation. The industrial restructuring of the US, currently under way, is commonly assumed to be driven, at least in part, by the diffusion and adoption of new technological capabilities. As a consequence, in the wake of such technological change many view the industrial landscape as being increasingly divided into 'hi-tech' and 'low-tech' production arrangements; the nation's economic landscape as being divided into 'prospering' and 'dying' regional and local economies; and workers as being segregated by whether or not their skills are appropriate to the range of tasks that will dominate and define the emerging new industrial economy. Most disturbing of all is the fear that these patterns will be highly correlated, resulting in older industrial sectors, areas and workers being locked in a mutually destructive embrace while industries and workers in regions elsewhere are destined to grow and prosper.

The trade-offs between enhanced productivity and international competitiveness and employment security accompany the passage from older industrial arrangements. However, while scientific advancements and technological innovations will surely figure prominently in the economic growth of new industries and new localized economies in the future, the adaptation of advanced design, production and related manufacturing technologies to older industrial arrangements likewise promises to revitalize from within the existing industrial base of the nation. Revitalized older industries (and perhaps the regions in which they are located) can be expected to take their place alongside wholly new industries in other regional economies to define a new national industrial structure. Older industries may be not so much jettisoned and discarded, as reorganized and thereby restored to higher, if not historical, levels of activity.

Our purpose in this chapter is to examine differences in the spread of key production innovations and their labour impact on selected machinery industries across the United States, leading to an examination of policy alternatives that would encourage further economic growth across the country. Just as the rate of technological change can be directly related to economic growth at the national level, so can the innovation level of states and metropolitan areas be related to their growth rates. Indeed, state and metropolitan differences in manufacturing productivity may be related to the failure of plants in some areas to adopt the latest production innovations.

This chapter will present the results of a research project that analysed the spread of key production innovations in manufacturing. In particular, this work examines the spread of major production innovations related to the use of automated machine control systems, the use of computers, programmable handling systems and microprocessors among machinery manufacturers across the United States. All these techniques relate to the degree of automation in manufacturing and will have substantial impacts on employment levels in the long run, in terms of both new and existing jobs. A questionnaire survey was sent out to nearly 4,000 manufacturing plants across the country and responses allow us to relate technology adoption rates to a number of variables: industrial sector, organizational type (single- and multi-plant firms), size and age of plant, the amount of research and development activity carried out, and locational characteristics of the plant.

2. Background

2.1 *The metalworking machinery industry as a setting for change*

To study change in the US metalworking industry is tantamount to studying this nation's adaptation to the urban–industrial era itself. The evolution of the metalworking industry over the past two centuries reflects not only the gradual domination of the US economy by manufacturing, but also the increased mechanization of production and its increasing dependence on metal products and tools required by basic manufacturing processes (Rosenberg, 1972).

Today, the 'metalworking industry' is multifaceted and the term serves as an 'umbrella' for a wide variety of disparate industries including the primary and fabricated metal industries, the machinery and electronic equipment industries, and the transportation and instruments industries, among others. While a common characteristic of all these industries is the fact that the medium for manufacture is metal, long years of product differentiation and process development have created an overall industry which is today highly variegated.

Yet within this seeming diversity lurk several enduring common characteristics which suggest an industry at large which has been slow to change. As the metalworking industry has matured, certain structural features have operated to constrain its technological revitalization. Since the industry possesses thousands of small plants and shops, process technologies have either not in the past been available to handle tasks at the appropriate small scale of such firms, or the productivity gains offered have not been sufficiently attractive in view of the required investment. Partly due to such small plants and shops, the introduction of new process technologies through the metalworking industry proceeded slowly until the 1970s. However, as the next section of this chapter shows, larger and older plants within the metalworking machinery industries have consistently shown a higher propensity to adopt new technologies, thus displaying a predisposition to retooling and rejuvenation.

2.2 *Advanced technology adoption in the metalworking industry*

A recent study by the Office of Technology Assessment (1983) reported that the manufacturing sector of the economy is poised to experience sweeping changes attributable to the adoption of programmable automated machine lines such as robots, computer-assisted planning, computer-aided design and computer-aided manufacturing (CAD/CAM). Yet though the harnessing of computers to manufacturing and design processes is not new, until recently the diffusion of such technological sophistication into plants and shops has been limited to those firms which are established and large enough to afford the typically heavy capital investment.

Today, due to lower purchase costs, the recent development of scaled-down turnkey systems of machine control, the present need to compensate for long-term shortages of skilled workers, competitive pressures in localized economies, and contractual stipulations that frequently require ever greater quality control, newer technologies are more attractive to thousands of plants for which they were previously considered inappropriate. With the availability of relatively inexpensive CAD/CAM systems, many new technologies are now within the reach of medium-sized and small manufacturing and engineering firms (Bylinsky, 1982). It is apparent that much advanced manufacturing technology which has been available in principle for many years is now filtering down to smaller firms in the metalworking industry after having long been hindered from doing so.

2.3 *The geographical patterns of industrial development*

The importance of the geographical concentration of the metalworking industry throughout the 'industrial heartland' and the diffusion of new technologies cannot be ignored (especially the East North Central and Mid-Atlantic Census divisions). From Table 7.1 we see that 42 per cent of all plants in the industry are located in these two regions that make up the traditional heartland of American manufacturing. Furthermore, the industry as a whole is tied into a dense network of suppliers, subcontractors and industrial customers, some of which are themselves highly localized. This pattern of concentration tends to amplify the consequences of intraregional competition and the job and market losses associated with industrial restructuring. Moreover, the industry remains sensitive to the relatively low value-to-weight ratios of its products that quickly translate into higher transportation costs between plants and shops which work with metal. In a sense, then, the locational immobility of the industry has matched, at least within small plants, a comparable intransigence towards the adoption of new production arrangements inside the plants concerned.

The industry appears to have spread slowly from its Mid-Atlantic and New England regions of origin. However, following World War II, the pace of industrial expansion from the East North Central region surpassed that found elsewhere, producing industrial development in a band of states from New York westward to Illinois, and this continued through the 1970s. The expansion within the Middle Atlantic region paralleled that of the East North Central until it began a relative decline after the 1960s.

Table 7.1 Geographical distribution of metalworking industry plants

	N	%
New England	790	9.8
Maine	25	
New Hampshire	60	
Vermont	14	
Massachusetts	390	
Rhode Island	37	
Connecticut	264	
Middle Atlantic	1,571	19.7
New York	596	
New Jersey	391	
Pennsylvania	584	
East North Central	2,574	32.2
Ohio	762	
Indiana	241	
Illinois	711	
Michigan	573	
Wisconsin	287	
West North Central	621	7.8
Minnesota	219	
Iowa	80	
Missouri	188	
North Dakota	4	
South Dakota	17	
Nebraska	30	
Kansas	83	
South Atlantic	546	6.8
Delaware	4	
Maryland	67	
Washington, DC	1	
Virginia	67	
West Virginia	24	
North Carolina	130	
South Carolina	55	
Georgia	48	
Florida	150	

	N	%
East South Central	193	2.4
Kentucky	63	
Tennessee	71	
Alabama	47	
Mississippi	12	
West South Central	481	6.0
Arkansas	29	
Louisiana	33	
Oklahoma	89	
Texas	330	
Mountain	221	2.8
Montana	1	
Idaho	3	
Wyoming	2	
Colorado	76	
New Mexico	19	
Arizona	73	
Utah	38	
Nevada	9	
Pacific	997	12.5
Alaska	1	
Washington	87	
Oregon	52	
California	857	
Hawaii	—	
Total	7,994	100.0

Source: Compiled from *Modern Machine Shop* (Cincinnati), *American Machinist* (New York), and *County Business Patterns* (US Bureau of Census).

The growth of the Far West (notably California) was largely a post-World War II phenomenon, yet it was eclipsed in each decade by that in the East North Central and Middle Atlantic regions. Finally, the growth in the South (i.e. South Atlantic, East South Central and West South Central) generally came later than that in the Far West. As the South gradually became a manufacturing centre (after 1960 eclipsing the West as the population growth pole of the nation), the metalworking industry likewise filtered into the South. Yet, like the West, the South has never seriously challenged the industrial heartland for dominance in the metalworking industry.

With regard to new firm formation, new plants in three age cohorts from 1950 to 1980 have largely occurred in the older industrial regions. Of the 207 new plants in the 1950s cohort, the 326 plants in the 1960s cohort and the 318 plants in the 1970s cohort, 57 per cent, 47 per cent, and 43 per cent respectively, were established in the East North Central and Middle Atlantic regions. Even during the recession-plagued 1970s and early 1980s, the older industrial regions continued to spawn new plants with 43 per cent of those in the 1980s being located in the East North Central region alone (Hicks, 1983).

Therefore, the Industrial Heartland appears to have more than held its own in terms of the revitalization of the metalworking industry through new plant growth. In these data there is little to support the common belief that older industrial regions are less susceptible to renewal and rejuvenation through new industrial growth and expansion. An industrial rejuvenation has indeed altered the structure of the metalworking industry, as this study discusses, but it has largely done so within its original industrial seedbed.

2.4 *Automation and employment change*

Both the machine tool industry and a whole host of satellite and interdependent industries (including the linked industrial sectors commonly referred to as the metalworking industry) have been faced with the need to adopt, rather than resist, the production process changes which promise to transform dramatically the nation's economy.

While concern has been shown towards the relatively slow growth in the manufacturing sector as a result of outdated production methods, it is also clear that the much needed improvements in efficiency afforded by automation will further dampen employment growth in manufacturing. It is in this context that the role of automation invites special concern. The adoption of ever more sophisticated technologies, and their diffusion throughout entire industries, is centuries old. The potential in

such circumstances for massive job displacement has periodically generated bursts of deep concern. The celebrated misgivings of the Luddites
about mechanized production in the British textile industry notwithstanding, there is historical evidence that, on balance, automation generally has been induced by the desire to maximize production efficiency,
rather than to displace labour *per se* (Ferguson, 1981). However, such
motives largely predated conditions in which skilled labour obtain relatively high wages, insulated from adjustment to changing market conditions by strong industrial unions.

It has been suggested that automation could bring about a major
reduction in the factory work-force over the next ten years. Yet in the
metalworking industry this general prophecy needs to be more carefully
qualified. Given the chronic shortage of skilled machinists available for
conventional machine control operations, a surge of automation is not
expected to generate as much job displacement in machining as in fabrication and assembly (Bylinksy, 1982). Nevertheless, it was not until the
recent recession and the forecast of widespread permanent job loss that
the full weight of concern about automation has been felt. For some,
the fall-off of investment in capital equipment since the 1960s appears
to belie any accelerated trend toward greater automation. Yet the manifestation of factory closures and the estimates that private disinvestment
by American business resulted in the loss of 30–40 million jobs in the
1970s alone have revived widespread anxiety over the role of automation
in industrial change (Bluestone and Harrison, 1982).

3. Research design

In the study discussed in this chapter, a discrete number of product and
process innovations within the manufacturing industry were selected as
the focus of investigation.[1] All the innovations relate, directly or indirectly,
to computerized automation within manufacturing and represent a set of
techniques at differing levels of sophistication that may have a significant
long-term impact on the American labour force and on productivity levels.
The innovations selected relate to four main areas of production technology: machine control, the use of computers, handling systems and the
use of microprocessors.

The specific techniques examined are:

- numerical machine control devices (NC);
- computerized numerical control devices (CNC);[2]
- computers used for commercial activities *only*, e.g. invoicing, stock
  control, accounting;

- computers used for design and drafting activities;
- computers used in manufacturing (excluding CNC);
- programmable handling systems for materials and subcomponents, including numerically controlled pick-up-and-place devices and simple programmable robots;
- non-programmable handling systems for materials and components, including manual and non-programmable pick-up-and-place devices;
- the use of microprocessors, mini- and micro-computers in the *final* product of a plant.

The first six production techniques relate directly to increased automation in the production process. Non-programmable material handling systems were included to isolate plants with more traditional handling devices. The use of microprocessors in the final product was the only product-orientated innovation examined.

The selection of innovations for study and the choice of industries as potential adopters were interrelated issues because the choice of innovation suggests particular sectors; for example, the use of NC and CNC suggests the metalworking machinery industry. Furthermore, to limit the scope of the study, and to facilitate interregional and international comparisons, it was necessary to delineate clearly a number of industries (by 3- and 4-digit SIC classification) as candidates for adopting the above innovations. The choice of a limited number of target sectors also acts as a control for industrial structure and indicates how it influences technology diffusion levels.

The six target industrial sectors chosen were producers of:

- farm machinery (SIC 3523);
- construction and related machinery, including elevators, conveyors, cranes and industrial tractors (SICs 3532, 3534, 3535, 3536, 3537);
- metalworking machinery for cutting and forming (SICs 3541, 3542);
- electrical distributing equipment, including transformers and switchgear (SICs 3612, 3613);
- electrical industrial apparatus, including motors, generators and welding equipment (SICs 3621, 3623);
- aircraft and parts, including engines (SICs 3721, 3724).

Most of the target population of potential adopters, amounting to 94 per cent of respondents, were machinery manufacturers (SICs 35 and 36). Thus the study was restricted to integral parts of the capital goods sector.

A postal questionnaire survey was sent to 3,873 individual manufacturing plants in the target sectors employing over twenty people as identified in the DUNS files of the Dun and Bradstreet Corporation (1976).[3] The questionnaire was sent out between February and April 1982 to all plants across the United States identified in the DUNS files as producing goods with the above SIC codes. This ensured extensive geographical coverage of the United States, as suggested in Table 7.2. Plants employing less than twenty people were omitted from the survey because past research has shown high death rates and lower response rates from this group.

A total of 628 completed questionnaires were returned. When undelivered questionnaires were discounted (either because the plant had moved to an unknown address or had gone out of business), the response represented an adjusted rate of 20 per cent. This level of postal questionnaire response is particularly good when compared with other studies of this kind and where success depends on the co-operation of busy corporate executives. Because a major purpose of this study was to examine differences in innovation adoption across a limited number of industrial sectors, it was particularly important that respondents to the mail survey represented a random geographical sample. A chi-square statistic of 13.12 showed no significant difference between the proportion of responses compared to the total population, i.e. the responses were random geographically.

Table 7.2 Potential adopters by industry and region

Sector	North East	North Central	South	West	United States
Agric. machin.	24	411	164	96	695
Machine tools	222	452	72	89	835
Construct. equip.	53	211	108	56	428
Mech. handling	156	357	153	117	783
Electrical machinery	234	354	177	125	890
Aircraft and parts	63	54	63	62	242
Total	752	1,839	737	545	3,873

Data Source: Dun and Bradstreet (plants with more than 20 employees)

4. Results

Tables 7.3 to 7.9 show the rates of adoption of the eight technologies according to the various characteristics of the manufacturing plants surveyed. Adoption rates (percentages) are displayed and chi-square tests of statistical significance[4] were performed on the absolute number of adopters per cell.

4.1 Adoption rates by industrial sector

Table 7.3 shows adoption or user rates by industrial sector, using the 3-digit SIC code of the US Census. Thus, of the 132 makers of agricultural machinery in Table 7.3, 20 per cent had adopted numerically controlled machines in their production process. When differences in adoption rates are analysed by industry, using a chi-square test, there are statistically significant differences in the adoption patterns, but only for five out of the eight technologies (Table 7.3). These differences are discussed below according to the four major groups of techniques surveyed.

(i) *The use of machine control systems.* The use of numerically controlled machinery varied from a 20 per cent adoption rate among producers of agricultural machinery to a 68 per cent adoption rate among aircraft manufacturers. The same general pattern is true for the use of computerized, numerically controlled machinery.

In four of the six industries, the adoption rate for CNC was higher than that for NC, suggesting that companies who had adopted NC also opted for the more advanced production technology. CNC is a major step in what Nelson and Winter (1977) call the natural trajectory of technological evolution from, in this case, manual control systems to advanced forms of automated production. The aircraft industry stands out as the major user of both NC and CNC largely because the Department of Defense, and the US Air Force in particular, has played a major role in the development of automated production through its programme for integrated computer-assisted manufacturing (ICAM) (National Research Council, 1981). The metalworking machinery industry exhibits adoption rates of over 50 per cent for both NC and CNC systems, probably because companies in this industry were the most directly involved in the manufacture of these products (Rosenberg, 1972).

(ii) *The use of computers.* When adoption rates for the use of computers for commercial activities are examined by sector, no statistically significant differences are evident. Adoption rates greater than 60 per cent of all

Table 7.3 Adoption rates by industrial sector

	Farm mach. (352)	Constr mach. (353)	Metal-work mach. (354)	Elec. distr. equip. (361)	Elec. indust. appar. (362)	Air-craft (392)	χ^2	Prob.†
NC	20	43	58	23	36	68	65.6	**0.0001**
CNC	23	37	58	27	44	70	54.6	**0.0001**
Computer for commercial use	63	69	61	67	62	82	6.9	0.228
Computer for design	10	21	19	36	28	51	36.6	**0.0001**
Computer for manufacture	34	49	46	41	40	55	8.7	0.122
Programmable handling devices	4	6	5	8	7	18	10.1	0.07
Non-programmable handling devices	47	45	36	48	46	68	14.2	**0.014**
Microprocessors in products	11	21	41	23	28	31	34.7	**0.0001**
Total no. of respondents*	132	170	152	77	57	40		

* The number of respondents is not necessarily the same for each technique due to a limited number of missing values.
† Statistically significant probabilities are shown in bold type.

plants are evident in all six industries, and reach 82 per cent in the aircraft industry. This is not an unexpected pattern, given that one might expect most companies today to use computers on site in their non-manufacturing activities for accounting, invoicing, or payroll functions.

In an examination of the use of computers for design, on the other hand, adoption rates are much lower and the difference between sectors is statistically significant. Again, the aircraft industry is the most innovative sector in its adoption of computers for design purposes (51 per cent), while the makers of farm machinery are the least innovative. The use of computers in the manufacturing process *per se* (excluding CNC) is more widespread than for design, but a statistically significant pattern is not evident between industries.

(iii) *Handling systems.* The rate of adoption of programmable or computerized handling systems is low in all sectors, with user rates below 10 per cent in five out of the six industries (the exception being aircraft). Because the development of robotic handling systems is still in its infancy, this pattern is expected. On the other hand, the use of non-programmable handling systems is more widespread throughout all the sectors in Table 7.3, with five out of the six showing adoption rates above 40 per cent.

(iv) *Use of microprocessors in final products.* The use of microprocessors as components in the final products of the plants surveyed (a product-as opposed to process-orientated innovation) shows statistically significant differences between sectors. The most innovative sector in this regard is the metalworking machine tools industry, which has increasingly used microprocessors in its products over time, as shown by the development of computerized numerical control systems by the industry. The second largest user of microprocessors is the aircraft companies, which use microprocessors, mini- and microcomputers in their instrumentation and control systems.

4.2 *Adoption rates by organizational status*

Table 7.4 shows adoption rates for each of the eight technologies under study according to the affiliation of the plants; whether they are part of a multi-plant firm (MPF) or a single-plant entity (SPF). A striking pattern emerges, which is both consistent for all the technologies and statistically significant in each case. Plants which are affiliated to multi-plant corporations have much higher rates of adoption than single-plant firms. For numerically controlled machines, the use of computers in design

Table 7.4 Adoption rates by organizational status

	SPF	MPF	χ^2	Prob.*
NC	25	56	58.8	**0.0001**
CNC	31	51	26.0	**0.0001**
Computer for commercial use	54	78	37.3	**0.0001**
Computer for design	11	34	39.4	**0.0001**
Computer for manufacture	29	57	44.2	**0.0001**
Programmable handling devices	2	11	23.6	**0.0001**
Non-programmable handling devices	39	51	9.9	**0.002**
Microprocessors in products	19	33	15.4	**0.0001**
Total no. of respondents	322	306		

* Statistically significant probabilities are shown in bold type.

and manufacturing, and for programmable handling systems, adoption rates among multi-plant companies are double what they are for single-plant companies. This may not be surprising when one considers the financial resources available to multi-plant firms, as suggested by the economies of scale implicit in such industrial enterprises.

This does show that multi-plant companies are more innovative in their introduction of new *process technologies* than single-plant companies. Although data on company size (as measured by total sales or assets) were not obtained directly in this survey, multi-plant companies are inevitably larger than single-plant firms. From Table 7.4, therefore, it can be inferred that larger multi-plant enterprises are more likely to adopt the latest available *process innovations* than are smaller, single-plant companies. It should be recalled, however, that small firms tend to introduce product rather than process innovations due to the reduced viability of high-volume process techniques (Utterback, 1979). These findings do, however, run contrary to the popularized notions that small, single-plant companies are relatively more innovative than their larger counterparts for all kinds of technologies, and they point out the importance of distinguishing between product and process innovations.

4.3 *Adoption rates by size of plant*

Although data were not collected on the size of overall corporations, the fact that the study was conducted at the level of the individual plant does allow us to address adoption rate differentials by employment size of plant. Again, a consistent and statistically significant pattern emerges for seven out of the eight technologies. As shown in Table 7.5, larger plants in the survey show consistently higher rates of innovation adoption than smaller plants. Table 7.5 uses the employment-size classification of the Economic Census, and shows consistently higher rates of adoption for all but one of the technologies as plant size increases from the twenty to ninety-nine employment-size category to plants employing 1,000 or more.

The increase in adoption rates for these technologies with an increase in the plant-size scale is highly consistent, ranging from 25 per cent adoption of NC in the twenty to ninety-nine employees category, to 83 per cent adoption for plants employing over 1,000. The only exception to this progression is the use of non-programmable handling systems. Higher adoption rates among smaller plants in this case is understandable when it is considered that this type of technology can include simple, automatic material-handling systems which are easier and cheaper to use in small plants. On a general level, however, it should be noted that these data on firm size strongly resemble plant status since small plants are more frequently independent, while large plants are often a member of the multi-plant group.

4.4 *Adoption rates by plant age*

The results of Table 7.6 show the least expected and perhaps the most provocative findings to emerge from this study. A priori we expected to find newer plants to be more innovative in their use of new technologies than older plants. Our findings, however, show the reverse to be the case, and this pattern is both consistent and statistically significant for six of the eight technologies. On the whole, older plants are more prevalent users of new process technologies than the new ones. For NC and CNC machine control systems, and for the use of computers in commercial, design and manufacturing activities, manufacturing plants built prior to 1939 show higher adoption rates than do plants built after 1940. Indeed, when plant age is compared by decade, a progressive inverse relationship exists between the age of plants and their propensity to adopt new technologies.

These results therefore suggest that, in a key part of the durable goods

Table 7.5 Adoption rates by size of plant (employees)

	1–19	20–99	100–249	250–999	1,000 or more	χ^2	Prob.*
NC	10	25	43	67	83	107.4	**0.0001**
CNC	8	23	50	69	78	121.2	**0.0001**
Computer for commercial use	24	50	77	91	95	114.7	**0.0001**
Computer for design	3	9	21	41	80	125	**0.0001**
Computer for manufacture	8	21	53	74	90	153.5	**0.0001**
Programmable handling devices	0	1	2	15	35	88.1	**0.0001**
Non-programmable handling devices	48	43	39	51	60	8.2	0.083
Microprocessors in products	5	19	32	36	40	29.2	**0.0001**
Total no. of respondents	40	279	135	125	40		

* Statistically significant probabilities are shown in bold type.

Table 7.6 Adoption rates by plant age

	1939 or before	1940–9	1950–9	1960–9	1970–81	χ^2	Prob.[†]
NC	59	52	41	33	28	32.7	**0.0001**
CNC	57	46	45	37	27	26.0	**0.0001**
Computer for commercial use	79	70	67	62	58	13.3	**0.009**
Computer for design	41	30	23	18	14	26.3	**0.0001**
Computer for manufacture	58	57	45	40	30	23.5	**0.0001**
Programmable handling devices	9	16	6	5	2	16.3	**0.003**
Non-programmable handling devices	34	49	49	48	46	6.7	0.150
Microprocessors in products	31	28	21	28	19	6.2	0.183
Total no. of respondents	111	63	109	181	150		

* Statistically significant probabilities are shown in bold type.

sector, older manufacturing plants across the country have been rejuvenating their process technologies in a drive to remain competitive. Much of this retooling can be explained by the fact that most of the new technologies are discrete units that can be introduced into a plant without great trauma. For example, a CNC system can be introduced into an existing plant for metal-cutting or metal-forming without a total reorganization of the plant's production layout. This is particularly true of computers used in commercial or design activities. The above results clearly imply that older plants in the United States cannot be written off as users of outdated technology. The results are also testimony to the inherent potential that older plants may have for increasing their technological sophistication.

One other explanation for the patterns evident in Table 7.6 lies in the consolidation or rationalization procedures that may have been experienced by some of the multi-plant companies surveyed. During times of recession or organizational restructuring it is possible that one or two newer plants within a multi-locational system may have been closed and the best available technology consolidated in an older plant at or near the original headquarters location of the firm. Yet this trend would have had to be a major one among most of the 628 respondents to account for the consistent patterns seen in Table 7.6.

The only exceptions to the patterns seen in Table 7.6 are for non-programmable handling systems and the use of microprocessors in final products, where no statistically significant differences in adoption rates are evident by plant age. Adoption rates for non-programmable handling systems do not vary much by plant age or size, since such systems are used by most plants. As for the use of microprocessors in final products, older plants are again relatively more innovative users than are the newer plants, but not to a statistically significant degree (Table 7.6).

The results of Table 7.6 do, however, point to the importance of differentiating between age of plant and age of capital stock when assessing the technological sophistication of US industry. Indeed, the potential among older plants for using the best available or practical process technologies can be directly related to the product-cycle argument for regional industrial change developed elsewhere (Rees, 1980; Erickson and Leinbach, 1979). Since most newer plants are likely to be branch plants, the product-cycle argument suggests that branch plants produce more mature products using standardized process technology. The standardization of production implies a lesser need to introduce more flexible processes like CNC, whose adaptability is better suited to the early types of product development in older plants, frequently located at or near the corporation's headquarters and central R & D.

4.5 *Adoption rates by the presence of research and development*

Table 7.7 examines variations in adoption rates according to whether research and development (R & D) activity is conducted in the manufacturing plants surveyed. This allows us to test whether or not the plants with on-site R & D are more likely to use new technologies. From Table 7.7 we see that 505 plants, or 80 per cent of the total, performed some form of R & D activity on site, while only eighty-seven plants, or 14 per cent of the total, had no R & D activity on site. Largely because of the high proportion of plants with R & D on site, no statistically significant differences in adoption rates were found for five out of the eight technologies when related to the presence or absence of R & D.

For users of computers in commercial activities, 70 per cent conducted R & D at the same location. For users of computers in the manufacturing process *per se*, 59 per cent conducted R & D at a separate location within the firm. Significant differences in adoption rates also emerge for users of microprocessors in their final products. This last pattern shows that

Table 7.7 Adoption rates by the incidence of R & D

	No R & D	R & D at another location	R & D on site	χ^2	Prob.*
NC	34	54	40	4.1	0.127
CNC	37	54	41	3.2	0.198
Computer for commercial use	44	59	70	21.3	**0.0001**
Computer for design	14	23	24	4.1	0.130
Computer for manufacture	23	59	46	16.4	**0.0003**
Programmable handling devices	4.6	14	6	3.8	0.153
Non-programmable handling devices	45	50	45	0.4	0.818
Microprocessors in products	15	12	28	9.8	**0.008**
Total no. of respondents	87	36	505		

* Statistically significant probabilities are shown in bold type.

the more innovative users of microprocessors in their final products had a substantial amount of R & D on site, a pattern that might be expected from the creative nature of such endeavours where much on-site work would have been needed to apply the microprocessors to existing or new products.

For five of the eight techniques, plants with R & D activities located at some other site within the corporate system showed the highest adoption rates. Because of the large number of respondents with R & D on site, adoption rates were also examined according to the number of R & D workers as a proportion of total employment at each plant. A table of results is not included here because the trends are very similar to those in Table 7.7. Only seventy-five plants (12 per cent of total respondents) had R & D workers that amounted to 5 per cent or more of total employment at that plant, while a mere twenty-one plants reported over 10 per cent of their workers as R & D personnel.

4.6 *Adoption rates by metropolitan location of plants*

Table 7.8 shows adoption rates according to the metropolitan character of the counties in which respondents are located. The fourfold division of counties in Table 7.8 includes:

- *large metro* implying counties within standard metropolitan statistical areas (SMSAs) of over 1 million people;
- *small metro* defined as counties within SMSAs of less than 1 million;
- *urban* implying non-metropolitan counties that include at least one city with over 10,000 population;
- and *rural* including non-metropolitan counties with *no* city over 10,000 people.

Table 7.8 shows statistically significant differences in adoption patterns for only two of the eight technologies: numerical control and the use of microprocessors in the final product. The adoption rate of NC is highest for plants in the *smaller SMSAs*, not the largest, while the lowest adoption rates occur in the rural areas. This same pattern is also true for plants using microprocessors in their final products. Indeed, adoption rates in the largest urban agglomerations are highest for only five of the eight technologies, and they are only marginally higher for two of these; CNC and non-programmable handling systems. This therefore suggests that the largest urban areas are not necessarily the most conducive environments for companies that use the latest available technologies. The adoption rates seen in Table 7.8 do suggest that smaller SMSAs, and to a large extent the more urbanized of the non-metropolitan counties, are also

Table 7.8 Adoption rates by metropolitan location

	Large metro	Small metro	Urban	Rural	χ^2	Prob.*
NC	43	46	36	30	8.7	**0.03**
CNC	43	42	41	33	3.06	0.383
Computer for commercial use	62	66	74	62	6.07	0.108
Computer for design	26	20	25	19	2.4	0.492
Computer for manufacture	46	39	49	40	3.95	0.267
Programmable handling devices	7	6	7	4	1.18	0.759
Non-programmable handling devices	44	43	48	46	0.879	0.831
Microprocessors in products	28	33	17	18	12.2	**0.007**
Total no. of respondents	218	175	140	95		
Average rank (exc. non-programmable handling)	1.7	2.3	2.1	3.6		

* Statistically significant probabilities are shown in bold type.

conducive environments for the adoption of these new production techniques. For three of the eight technologies (computers for commercial and manufacturing activities and non-programmable handling systems) the more urbanized non-metro counties show the highest adoption rates. Although the larger SMSAs still show the highest average ranking for all technologies, bar non-programmable handling, the more urbanized non-metro areas show the second highest ranking, followed by the smaller SMSAs and then the more rural areas.

4.7 *Adoption rates by region*

One of the major goals of the research was to examine differences in innovation adoption by geographical region, based on the hypothesis that plants in various parts of the country might show variations in their propensity to adopt the latest technology. Table 7.9 shows variations in adoption rates by Census region, based on a random response pattern. Although statistically significant differences in adoption rates only appear

Table 7.9 Adoption rates by census region

	North East	North Central	South	West	χ^2	Prob.*
NC	39	45	32	35	7.68	0.053
CNC	41	47	28	37	12.4	**0.006**
Computer for commercial use	62	69	63	62	2.7	0.441
Computer for design	23	22	23	25	0.2	0.977
Computer for manufacture	47	46	38	36	3.9	0.272
Programmable handling devices	6	7	4	11	3.9	0.267
Non-programmable handling devices	40	42	55	51	8.4	**0.038**
Microprocessors in products	31	26	20	23	4.3	0.226
Total no. of respondents	114	325	128	61		
Average rank (excl. non-programmable handling)	2	1.9	3.3	2.6		

* Statistically significant probabilities are shown in bold type.

for two of the eight technologies, there are some important regional differences in the adoption rates for the survey innovations.

Regional differences in the adoption of CNC are statistically significant, with the North Central region showing an adoption rate of 47 per cent, followed by the North East, the West and the South. The high adoption rate for CNC in the North Central region may be expected from the region's industrial base which includes the largest industrial states of the Manufacturing Belt (Michigan, Ohio, Illinois) and the area's role as the historic centre for the machine tools industry (Rosenberg, 1972). The North Central region also has the highest adoption rate of NC, where the adoption pattern by region is similar to that for CNC, for reasons similar to those above. The North Central region also shows the highest adoption rate for the use of computers for commercial activities.

In the case of computers for commercial activities, however, regional variations in adoption rates are very small. Since the use of computers for commercial purposes did not show statistically significant differences by sector (Table 7.3), it is not surprising that major differences do not emerge. Plants in all four regions of the United States show adoption rates above 60 per cent for the use of computers in commercial activities. It is perhaps more surprising that regional differences in the use of computers for design purposes, as well as for manufacturing, are not larger.

Adoption rates for programmable (mostly robotic) handling systems are low by region, as they are by sector. Regional variations in the use of non-programmable handling systems, on the other hand, are distinct and statistically significant. In this case it is the Southern region which shows the highest user rate and the North Eastern states the lowest rate. The high adoption rate in the South is testimony to the continued dominance of the region by branch plants based on lower technology standardized production (Hansen, 1980), despite the rapid growth of certain high-technology growth centres in the Sun Belt states (Rees, 1979). Regional differences in the use of microprocessors in final products are not statistically significant. The marginal dominance of the North East in this case is testimony in part to the development of mini- and microcomputers in areas within New England, such as Boston (Dorfman, 1982).

Given the size and diversity of the United States, it may not be surprising that a complex pattern of regional differences in the adoption of new technologies is indicated in Table 7.9. When an average ranking of regional adoption rates is carried out for seven of the eight technologies (non-programmable handling systems are left out because of their lower technology base), the dominance of the Manufacturing Belt as a user of the latest available process technology is evident. The North Central region ranks highest, followed by the North East, the West and the South. Although such rankings should not be overemphasized, they do indicate that, despite the relative growth of the South and West in the last fifteen years, industries in the growth regions are poorer adopters of the latest available technology. Indeed, as suggested by the age of plant variable in Table 7.6, it is the older industrial regions of the North Central and North Eastern parts of the Manufacturing Belt that display the highest propensity to use new production technologies.

(i) *The influence of organizational status.* When regional adoption rates are examined by organizational status (Table 7.10), statistically significant differences are evident between regions for *single-plant firms* adopting three key technologies: NC, CNC and microprocessors in the final product. These findings are important in that they show small, single-plant firms in

Table 7.10 Regional adoption rates by organizational status

		North East	North Central	South	West	Prob.*
NC	SPF	27	31	11	17	**0.02***
	MPF	55	60	49	53	0.47
CNC	SPF	37	37	16	13	**0.004***
	MPF	47	56	38	60	0.06
Computer for commercial use	SPF	54	58	47	43	0.36
	MPF	70	80	76	80	0.52
Computer for design	SPF	17	9	10	15	0.49
	MPF	31	36	32	34	0.93
Computer for manufacture	SPF	38	32	20	15	0.07
	MPF	57	60	52	55	0.76
Programmable handling devices	SPF	3	2	0	3	0.37†
	MPF	10	12	7	19	0.35
Non-programmable handling devices	SPF	37	35	47	47	0.30
	MPF	43	49	61	55	0.22
Microprocessors in products	SPF	33	16	11	20	**0.01***
	MPF	29	38	27	27	0.33

* Statistically significant (using chi-square) probabilities are in bold type.
† More than 20 per cent of cells have expected counts less than 5.

the industrial heartland (the North East and North Central regions) to have far greater adoption rates for NC and CNC than similar firms in the Southern and Western Census regions. Likewise the use of microprocessors in final products is more prevalent in single-plant firms in the North East and Western regions than in the Mid-West or South. It is no coincidence that in the case of CNC, most of the early development work was spawned in the Manufacturing Belt, whereas in the case of microprocessors in products, firms in Massachusetts and California appear to have been the most progressive in the development of mini- and microcomputers. For single-plant firms, therefore, this suggests a distance–decay effect in adoption patterns where adoption rates are lower in regions furthest removed from the origins of these leading-edge technologies. Because of the comparative advantage that multi-plant firms have in spreading new production technologies in a variety of locations within their corporate system, it is not surprising that multi-plant firms in Table 7.10 show much less regional variations in adoption rates for all the technologies studied.

The distance–decay effect for single-plant firms does not appear as statistically significant, however, when metropolitan and non-metropolitan adoption rates are compared in Table 7.11. Adoption rates for NC and microprocessors ae higher for plants in metropolitan areas than in non-metropolitan counties. Table 7.11 also shows adoption rates for NC and microprocessors are higher for plants in metropolitan areas than in non-*plant firms*, showing that these key technologies are more likely to be introduced in urban rather than rural plants of multi-locational firms. Presumably the more sophisticated labour force associated with urban rather than rural locations would be a major factor in the introduction of these relatively complex technologies.

(ii) *The influence of plant size.* Table 7.12 shows regional adoption rates by size of plants, using employment levels of below 100 workers to define smaller plants and employment levels of 100 workers or more to define larger plants. Regional adoption rates are not significantly different for any of the techniques, except CNC among the smaller plants. For the

Table 7.11 Metropolitan adoption rates by organizational status

		Large metro	Small metro	Urban	Rural	Prob.*
NC	SPF	27	29	22	19	0.56
	MPF	62	62	49	40	**0.03**
CNC	SPF	32	32	36	19	0.27
	MPF	56	52	46	47	0.58
Computer for commercial use	SPF	51	53	65	45	0.15
	MPF	74	79	83	77	0.56
Computer for design	SPF	12	9	13	12	0.83
	MPF	39	31	36	26	0.47
Computer for manufacture	SPF	32	27	36	17	0.17
	MPF	59	49	62	62	0.39
Programmable handling devices	SPF	1	4	3	0	0.36†
	MPF	15	9	12	9	0.54
Non-programmable handling devices	SPF	38	39	40	40	0.99
	MPF	51	48	56	53	0.77
Microprocessors in products	SPF	17	29	12	15	**0.04**
	MPF	41	36	23	22	**0.04**

* Statistically significant (using chi-square test) probabilities are shown in bold type.
† More than 20 per cent of cells have expected counts less than 5.

Table 7.12 Regional adoption rates by employment size of plant

		North East	North Central	South	West	Prob.*
NC	1–99	27	26	16	19	0.43
	≥100	54	63	50	54	0.27
CNC	1–99	29	24	9	16	**0.02**
	≥100	56	67	52	62	0.16
Computer for commercial use	1–99	46	48	46	39	0.83
	≥100	79	87	85	89	0.59
Computer for design	1–99	13	7	7	10	0.57
	≥100	37	35	43	46	0.66
Computer for manufacture	1–99	24	21	16	7	0.25
	≥100	69	68	65	68	0.97
Programmable handling devices	1–99	2	1	0	3	0.59
	≥100	11	12	9	19	0.65
Non-programmable handling devices	1–99	40	40	54	47	0.21
	≥100	39	44	55	59	0.16
Microprocessors in products	1–99	21	17	10	23	0.30
	≥100	42	35	31	26	0.48

* Statistically significant probabilities are shown in bold type.

smaller plants using CNC, however, adoption rates in the industrial heartland (the North East and North Central regions) are significantly higher than in the South and West. This suggests that the argument made earlier regarding single-plant independent firms also pertains to smaller plants since the bulk of small plants are independent firms. Regional differences in the adoption rate of small plants are also evident for NC and microprocessors, but are not statistically significant (Table 7.12).

(iii) *Differences due to plant age.* Because of the significant trends portrayed by the plant age variable at the national level in Table 7.6, regional and metropolitan differences in this variable are further explored in Tables 7.12 and 7.13. Here a dichotomous variable is used to define either those older plants established before 1960 or newer plants founded in 1960 or later. From Table 7.12 significant regional differences in adoption rates are evident for older plants using NC and CNC. Again, the role of the North Eastern and Mid-Western states as the fount of machine-tools technology is evident, with adoption rates among pre-1960 plants being higher in the North Central region than in the South. However,

Table 7.13 Regional adoption rates by plant age

		North East	North Central	South	West	Pron.*
NC	pre-1960	40	60	36	41	**0.005**
	1960 or later	39	29	29	29	0.51
CNC	pre-1960	43	60	29	41	**0.001**
	1960 or later	39	32	28	35	0.52
Computer for commercial use	pre-1960	71	77	67	54	0.08
	1960 or later	53	61	61	66	0.65
Computer for design	pre-1960	26	32	36	30	0.84
	1960 or later	20	13	16	23	0.40
Computer for manufacture	pre-1960	56	58	42	33	0.08
	1960 or later	38	34	36	37	0.95
Programmable handling devices	pre-1960	7	11	5	15	0.48
	1960 or later	5	3	4	6	0.66
Non-programmable handling devices	pre-1960	37	39	57	59	**0.04**
	1960 or later	42	46	53	44	0.58
Microprocessors in products	pre-1960	27	29	23	12	0.26
	1960 or later	33	22	18	31	0.17

* Statistically significant probabilities are shown in bold type.

Table 7.14 Metropolitan adoption rates by plant age

		Large metro	Small metro	Urban	Rural	Prob.*
NC	pre-1960	51	51	55	37	0.99
	post-1960	35	40	16	27	**0.009**
CNC	pre-1960	51	50	48	50	0.99
	post-1960	35	34	32	25	0.59
Computer for	pre-1960	66	72	76	85	0.24
commercial use	post-1960	57	59	71	53	0.17
Computer for	pre-1960	35	26	38	19	0.25
design	post-1960	17	14	13	19	0.75
Computer for	pre-1960	56	44	60	48	0.24
manufacture	post-1960	35	33	38	37	0.94
Programmable	pre-1960	14	6	10	3	0.21
handling devices	post-1960	1	6	4	5	0.8†
Non-programmable	pre-1960	46	42	45	33	0.67
handling devices	post-1960	42	45	51	53	0.48
Microprocessors	pre-1960	30	27	20	28	0.59
in products	post-1960	20	36	15	15	**0.007**

* Statistically significant (using chi-square) probabilities are shown in bold type.
† More than 20 per cent of cells have expected counts less than 5.

regional differences in the adoption of these technologies do not appear as statistically significant for plants set up after 1960.

User rates for non-programmable handling equipment also reveal statistically significant differences for older plants, showing the plants of the South and West to be the most frequent users. This reflects the more traditional handling systems that one may expect among the branch plants of peripheral regions in the South and West.

When adoption rates for older and newer plants are examined by their urban and rural locations (Table 7.13), the only statistically significant differences appear for newer plants introducing two innovations: numerical control, and microprocessors in products. Again these newer technologies are more likely to be introduced in the more sophisticated labour markets of metropolitan areas rather than non-metropolitan locations (Table 7.14). Unexpectedly in these cases, the same pattern does not hold for the older plants.

5. Conclusion

From this study of the spread of automated production technology in the American machinery industry we have seen that adoption rates do vary significantly by type of industry, by type of company, by size and age of plant and by the presence or absence of R & D. Our finding that older plants are more likely users of these new production technologies than newer plants is testimony to the technological change occurring in the more established industrial areas of the country. This rejuvenation process has been 'glossed over' by many recent studies of industrial change in the United States.

At its simplest, the above study gives evidence that market mechanisms are working in the sense that such retooling is mandatory for firms to remain competitive. Since these adoption patterns also reveal regional differences of varying significance, the overall pattern of results indicates a correlation between capital and labour by region. This relationship suggests that the more advanced production technologies are introduced in the higher skill, higher wage areas of the industrial Mid-West to yield cost-savings, while a lower incidence of these technologies (or less advanced versions) are found in the lower wage, lower skill labour markets of the South and West. Indeed, this alignment process can be seen to follow a product-cycle interpretation of regional industrial change proposed earlier for the United States (Rees, 1980). For example, the greater use of CNC in the industrial Mid-West suggests, at least for the machinery industry, that early development work is still ongoing in that region, while more standardized production is more prevalent in the peripheral regions of the South and West.

The policy implications in our findings are found at the regional scale, where small single-plant firms show significant differences in their propensity to adopt leading-edge technologies in two important instances. First, single-plant firms show far higher adoption rates for computerized machine control equipment in the industrial Mid-West, the spawning-ground for the initial development of this technology. Second, the use of microprocessors in final products is more prevalent in their region of origin: in this case the North East (notably Massachusetts) and the West (notably California). This suggests a contagious diffusion or distance-decay effect emanating from regions that spawn leading-edge technologies, and is testimony to the propulsive nature of innovative regions. Although (as might be expected) multi-plant firms show much less regional variation in the adoption of the technologies under study, they are clearly more prevalent users of key technologies (computerized machine control and microprocessors) in metropolitan rather than non-metropolitan environments. This again reflects the produce-cycle argument at the metropolitan scale (Erickson and Leinbach, 1979). For policy-makers interested in the nurturing of small business in particular, this study shows that small firms within the source region of leading-edge innovations are more likely to adopt such local new technologies. Hence some attention may need to be given to encouraging the spread of these technologies to less innovative environments, where multi-plant firms have a clear advantage over single-plant firms who suffer from the tyranny of distance.

Notes

1. This part of the paper draws upon a study by J. Rees, R. Briggs and R. Oakey, *The Adoption of New Technology in the American Machinery Industry*, Discussion Paper, Syracuse University, based on a project funded by the National Science Foundation's NSF Grant SES 810588. This is part of an international collaborative research project involving the United Kingdom and the Federal Republic of Germany under the co-ordination of Professor John Goddard.
2. NC machines are controlled by programs expressed in numbers, and are predecessors (on the road to fully flexible automation in manufacturing) of the more flexible and versatile CNC systems which are the equivalent of NC machines equipped with reprogrammable computers.
3. Though the accuracy of Dun and Bradstreet data has been questioned in studies of job creation, it remains the best national directory of manufacturing establishments available on computer tape.
4. The chi-square test is one of the most common analytical comparisons applied to multiple groups of data classified as frequencies. The result tests whether the observed frequencies of a given phenomenon (in this case adopters of particular innovations) differ significantly from the frequencies which might be expected (in this case from the general distribution of industry).

194   J. Rees, R. Briggs and D. Hicks

References

Bluestone, B. and Harrison, B. (1982). *The Deindustrialization of America* (New York, Basic Books).

Bylinsky, W. M. (1982). 'Computerized Design Systems Being Made for Smaller Firms', *Wall Street Journal* (12 November).

Dorfman, N. A. (1982). 'Massachusetts' High Technology Boom in Perspective', Center for Policy Alternatives, Discussion Paper, MIT.

Erickson, R. A. and Leinbach, T. R. (1979). 'Characteristics of branch plants attracted to nonmetro areas', in R. Lonsdale and H. Seyler, (eds) *Nonmetropolitan Industrialization* (Washington, DC, Winston), pp. 57–78.

Ferguson, E. S. (1981). 'History and Historiography', in O. Mayr and R. C. Post (eds), *Yankee Enterprise: The Rise of the American System of Manufactures* (Washington, DC, Smithsonian Institute Press).

Hansen, N. (1980). 'The New International Division of Labor and Manufacturing Decentralization in the United States', *Review of Regional Studies*, 9, 1–11.

Hicks, D. A. (1983). 'Technology Succession and Industrial Renewal in the U.S. Metalworking Industry', Discussion Paper, University of Texas at Dallas.

Nelson, R. and Winter, S. (1977). 'In Search of a Useful Theory of Innovation', *Research Policy*, 6, 36–76.

Office of Technology Assessment (1983). *Automation and the Workplace: Technical Memorandum* (Washington, DC, USGPO).

Rees, J. (1979). 'Technological change and regional shifts in American Manufacturing', *Prof. Geog.*, 31, 45–54.

Rees, J. (1980). 'Government Policy and Industrial Location in the United States', *Special Study on Economic Change*, vol. 7, Joint Economic Committee, US Congress, Washington, DC.

Rees, J., Briggs, R., and Oakey, R. (1983). 'The Adoption of New Technology in the American Machinery Industry', Discussion Paper, Syracuse University.

Rosenberg, N. (1972). *Technology and American Economic Growth* (New York, Harper and Row).

Utterback, J. M. (1979). 'The Dynamics of Product and Process Innovation in Industry', in C. Hill and J. Utterback (eds), *Technological Innovation for a Dynamic Economy* (New York, Pergamon).

8 The Dissemination of Public Sector Innovations with Relevance to Regional Change in the United States

JANET E. KODRAS
Florida State University, Tallahassee, Florida
and
LAWRENCE A. BROWN
Ohio State University, Columbus, Ohio

1. Introduction

Examinations of the role played by innovation diffusion in regional development have traditionally focused upon change in the secondary sector, such as shifts in industrial technology. However, the increasing prominence of the service sector in national economies indicates a further need to study its growing impact upon regional prosperity. The diffusion of a full range of service functions, from establishment of military installations to fast food franchises, has ramifications for place-specific prosperity, at different spatial scales. This chapter examines one segment of the service sector; innovations disseminated by the state in the form of government programs. Attention first turns to contemporary diffusion theory and its applicability in this context. By way of illustration, the second section presents a case study of the United States Food Stamp Program. The analysis demonstrates substantial spatial variations in program effectiveness, sources of that variation attributable to adopters and providers, and the ways in which diffusion theory may be applied to help explain the impact of this class of mechanism.

2. Application of diffusion theory to the public sector

2.1 *Perspectives on diffusion*

The dominant tradition of diffusion research has been the *adoption perspective* (Rogers, 1969; Rogers and Shoemaker, 1971), which focuses upon the potential adopter. One concern is awareness of the innovation, which is related to information dissemination and congruence between the mode by which that occurs and the individual attributes of potential adopters. A second concern is receptivity or resistance to adoption, which is a function of need for the innovation and/or the individual's innovativeness, a social-psychological trait.

In focusing upon aspects of *demand*, the *adoption* perspective tends to 'short-change' conditions beyond the individual's control which affect accessibility to, or availability of, the innovation, a prerequisite of the decision to adopt. This dimension is the primary concern of the *market and infrastructure* perspective (Brown, 1981). In emphasizing aspects of *supply*, one concern is the establishment of the diffusion agency, to provide outlets through which the innovation is distributed to the population at large. A second concern is the strategy implemented by each agency to induce adoption in its service area. From this perspective, adoption, the focus of most previous research, is closely linked to the diffusion agency establishment and dissemination strategy.

Considering the *adoption* and the *market and infrastructure perspectives* together, it is evident that unless the innovation is made available at or near the location of the potential adopter, through the establishment of a diffusion agency, the individual will not have the option to adopt in the first place. However, notwithstanding the presence of a diffusion agency, dissemination strategies may focus on some segments of the population, in lieu of others. Even in this case, access to the innovation does not in itself guarantee adoption. The innovation may not be congruent with the needs of the potential adopters or, alternatively, their psychological disposition toward acceptance may not be present.

To illustrate the interplay between supply and demand forces, consider the diffusion of cable television, a service sector innovation, throughout the United States (Brown, Malecki, Gross, Shrestha, and Semple, 1974). Federal communication regulations of the early 1950s prevented small towns and peripheral locations from receiving *broadcast* television. This supply constraint created a need for *cable* television as an alternative, and the initial distribution of cable systems reflected the spatially variable need. However, not all potential locales were serviced by a diffusion agency, since early systems were established through local initiative. Within each cable system's hinterland, furthermore, the diffusion strategy involved provision of cable to some neighborhoods prior to others, thus controlling the potential order of household adoption. Finally, some households availed themselves of this opportunity immediately, some hesitated, and others chose never to adopt. More recently, the function of cable television has shifted, from overcoming poor reception in remote areas, to increasing viewer options, and government regulations have been altered accordingly. As a result, corporate-owned diffusion agencies are more prevalent, their diffusion strategy is more sophisticated, and household need/demand for cable has changed.

2.2 *Public sector innovations*

Paralleling the larger body of diffusion research, work dealing with public sector programs has tended to take an *adoption perspective*. For example, the 'American states' literature (Walker, 1969; Gray, 1973) attributes differences in the timing of policy initiation by states to their relative innovativeness. Similarly, cross-national diffusion studies (Collier and Messick, 1975; Ross and Homer, 1976) attribute adoption of new programs, such as social security, to national innovativeness.[1]

Drawing upon the preceding discussion, we contend that the study of public sector innovations requires the combined strengths of the adoption and the market and infrastructure perspectives. In the case of entitlement programs, of which the United States Food Stamp Program is an example, at least four aspects of diffusion require examination:

1. the motivation underlying program availability;
2. the character and structural organization of government entities, serving as diffusion agencies;
3. the strategies by which the program is diffused;
4. the need/demand for government assistance on the part of the population.

The term *motivation* refers to the philosophy and broad guiding principles underlying the initiation, maintenance, and alteration of government programs. Whereas private sector innovations are generally driven by profit criteria, providing a single goal with a fairly clear-cut set of rules, public innovations often entail a complex set of purposes which may, in fact, be contradictory. Although the stated intention of policy may be service provision, the set of underlying objectives may differ considerably. As we shall discuss later, the stated purpose of the food stamp program was to assist low-income households in the purchase of a basic need, while the underlying intention was to assist agricultural interests by increasing demand for domestic food products. More generally, there are always differing attitudes about the rights and responsibilities of the state to provide public assistance and the appropriateness of different types of social and economic aid. It is necessary, therefore, to study the variety of actors at all levels of government bureaucracy, whose attitudes toward public assistance in general, and the individual program specifically, influence its design, organization, operation, and effectiveness.

One point at which these differing attitudes coalesce is in the character and organizational structure of the *diffusion agency*. In public sector diffusions, the structure, or control, of these entities often is decentralized, with the national government acting as overall coordinator and state/local units responsible for day-to-day administration.[2] Accordingly, there is considerable latitude for national, regional, and local interests to prevail differentially, and thus the nature of the program may vary considerably from level to level of the bureaucracy and, within each level, from place to place. This is illustrated by the location of diffusion agencies. According to service provision criteria, the order with which these are established should reflect the degree of need among the population. In fact, however, this concern may be overridden by broader political interests. Further, the local character of diffusion agencies may vary considerably, even if blanket coverage is mandated. Food Stamp offices, for example, are found in every US county, but they differ in locational proximity to the eligible population, size and quality of staff, convenience in application procedures, and so forth.

Differences in interpretation of the program's purpose are also manifest in the agency *dissemination strategies*. Broad outlines of this are often specified at high levels of government and might include decision-making powers associated with each level of the program's bureaucracy, such as market selection criteria based on whom the program is designed to serve, and the means of targeting that market, such as promotional campaigns, service infrastructures, and user costs.[3] Once again, however, local attitudes will affect implementation, particularly because subnational units are often given considerable discretion in program administration. For example, with regard to voter registration and school desegregation, local jurisdictions in the southern United States resisted new program guidelines, long after the programs had been established. Even more egregious examples are found in programs where eligibility criteria are left largely to local or state governments, as in the Aid to Families with Dependent Children program (Wohlenberg, 1976a, b).

The preceding aspects of entitlement program availability not only reflect *need/demand* for assistance, as indicated in several examples above, but also influence it. Potential user awareness of entitlement and willingness to respond accordingly are affected by diffusion agency actions and local, regional, and national attitudes toward public assistance. Illustrative of the latter is the moral authority of the US presidency, which has represented social welfare programs as either a haven for fraud or, alternatively, as a safety net for the deserving poor, and thus has profoundly affected societal values.

Even when a program is effectively supplied, however, *individuals* may nevertheless resist. Examples include opposition to early collectivization schemes in the Soviet Union and the Prohibition era in the United States, both of which ran counter to deep-seated values and traditions. Accordingly, a 'cultural fit' between public programs and potential users is an additional element of importance (Foster, 1962).

In summary, diffusion of public sector innovations occurs when the state, a complexly structured diffusion agency working with a variety of motivations and purposes, devises a set of strategies to provide the program; and when potential adopters decide whether or not to accept the program. These events are dependent upon the needs of potential adopters, the extent to which the policy identifies with their attitudes, and the influence of societal norms.

3. The United States Food Stamp Program

To illustrate further the applicability of diffusion theory to public sector innovations, attention now turns to the US Food Stamp Program (FSP). Presented first is a historical account which identifies supply, need, and demand factors influencing the program's operation and effectiveness at the national, state, and local levels. This is followed by statistical analyses, which examine the relative, and shifting, role of these factors at the local level. This case study demonstrates that the *market and infrastructure perspective* and the more commonly utilized *adoption perspective* are both essential for understanding diffusion of public sector programs.

3.1 *Historical account*

The US Food Stamp Program was established in 1964, and is currently one of the largest national programs of public assistance to the low-income population, both in terms of the number of people served and government expenditures. Participants receive food stamp coupons from the government, which are used in place of cash, to purchase domestic food products in local grocery stores (Allen, 1977; MacDonald, 1977). The benefit level varies by households such that those with the lowest income and of the largest size receive the greatest allotments. All benefit payments and approximately half the administrative costs are funded by the federal government.

The Food Stamp Program was framed in the rhetoric of anti-poverty legislation, but its original motivation was, in fact, to increase demand for domestic agricultural products. Anti-poverty forces had petitioned the government in every session of Congress since 1943 to institute a food stamp plan, but not until the President and the Secretary of the Department of Agriculture supported the issue was the proposal successful. That the Department of Agriculture, whose primary responsibility is to serve farm interests, was given responsibility for administration suggests that the program was oriented toward the agricultural sector. More specifically, the goal of increasing consumption of farm products was implied by the stipulation that low-income households were originally required to pay up to one-third of their income to participate. They received food stamps, of a higher monetary value, in return. By requiring that such a larger portion of low-income household budgets be committed to food purchases, the program thus favored agricultural interests over those of the poor. In fact, this particular provision is generally regarded as the major constraint which prevented many legally eligible recipients from participating.

Further controls on program participation were introduced at the subnational levels since state and local welfare offices were responsible for day-to-day administration. States originally had the option whether or not to implement the program and decided which counties could participate. By 1969, five years after FSP initiation, the program was operating in fewer than one-half the nation's counties (US Senate, 1969). Although the FSP was ostensibly a nationwide entitlement policy, the poor who happened to reside in any one of the 1,500 counties which did not operate the program were effectively not supplied. Furthermore, even where programs were available, participation levels varied substantially due to state control of eligibility criteria, which ranged from strict standards in the deep South and interior agricultural states to liberal criteria in urban industrial regions (MacDonald, 1977). Diffusion of the FSP was also influenced at the lowest level of the government heirarchy. Local welfare offices exerted considerable control in terms of application procedure stringency, office location, and hours of operation. Local governments also succumbed to negative community pressures; food stamp offices in several rural Mississippi counties, for example, were closed during harvest so that local farmers had access to a motivated labor force (James, 1972; Thorkelson, 1969).

A series of legislative revisions since 1964 altered the nature of the FSP, impacting upon participation levels and removing the basis for some of the inequities cited above. In 1971, nationally uniform eligibility standards were established; in 1974, all counties were required to provide the program; and beginning in 1979, households were no longer required to pay for food stamp coupons. Changes in individual need and demand for the program also influenced participation levels. A generally declining economy since the late 1960s, exacerbated by sharply rising food prices, increased the need for food stamps. The late 1960s and early 1970s also was a time of change in societal attitudes concerning public assistance;

in particular, welfare rights organizations were instrumental in decreasing the stigma attached to the use of welfare (Piven and Cloward, 1971; West, 1981).

To summarize the above, Food Stamp Program availability initially was constrained by federal guidelines pertaining to coupon payments, by state actions which made the program available in some counties but not others, and the local administrative practices. Many of these original constraints have since been altered through legislative revisions. In addition, a fluctuating economy and changing societal attitudes toward public assistance have altered the environment in which the program has operated.

Temporal and spatial variations in program use reflect these characteristics of, and changes in, supply, need and demand factors. Specifically, national participation levels have followed an essentially logistic growth trend from 1969 to 1980 (Fig. 8.1). The participation rate, defined as the proportion of the eligible population which actually takes food stamp assistance, was only approximately 10 percent in 1969; between 1969 and 1975, program use increased dramatically; between 1975 and 1979, participation rates fluctuated around 50 percent (Kodras, 1982). During this time period, interstate disparities in program use declined (Fig. 8.2). The coefficient of variability for 1969 was an extremely high 89 percent, but as legislative revisions loosened supply constraints, as awareness of the program increased, and as the stigma attached to welfare use declined, the coefficient decreased to 33 percent by 1975. It has remained at approximately this level as of 1979.[4] The following analyses seek to identify forces which have accounted for shifting disparities in program use at the local level.

3.2 *Statistical analyses*

Attention now turns to FSP effectiveness among the eighty-eight counties of Ohio for the years 1969, just prior to the program's major growth phase nationally, 1975, the end of that phase, and 1979, the most recent date for which data are available. This set of observations includes a variety of cultural and institutional settings: major metropolitan areas such as Cleveland, Cincinnati, and Columbus; historically poor, rural areas in Ohio's south-eastern Appalachian Mountains region; wealthy farming areas in the north-west; and long-standing industrial areas along the Ohio River, in the south, and along Lake Erie, to the north (Fig. 8.3).

FSP effectiveness is measured as the participation rate for county i at time t (PR_{it}), defined as the proportion of eligible persons who use the program.[5] These rates have varied substantially throughout the state

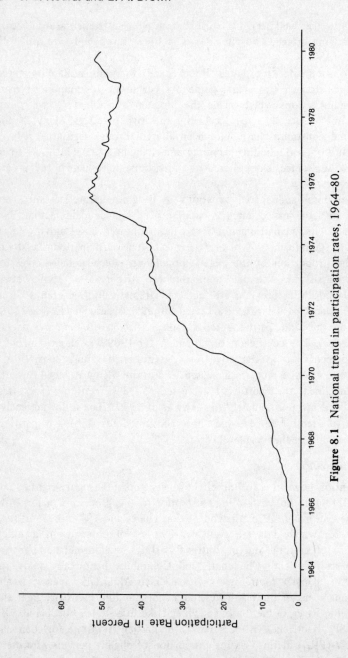

Figure 8.1 National trend in participation rates, 1964–80.

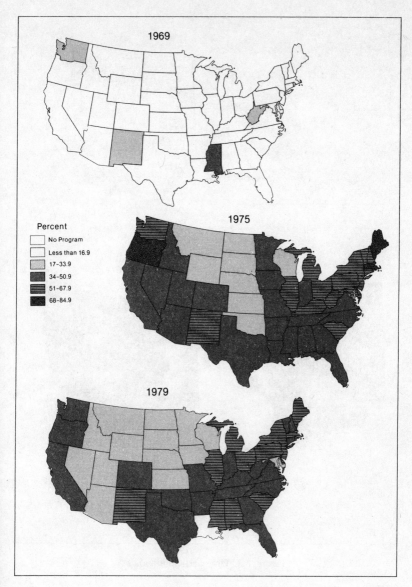

Figure 8.2. State participation rates

although disparities have declined through time (Fig. 8.3). Specifically, the 1969 coefficient of variability was 79 percent, when eighteen counties did not provide food stamps. This decreased to 29 percent by 1975 and

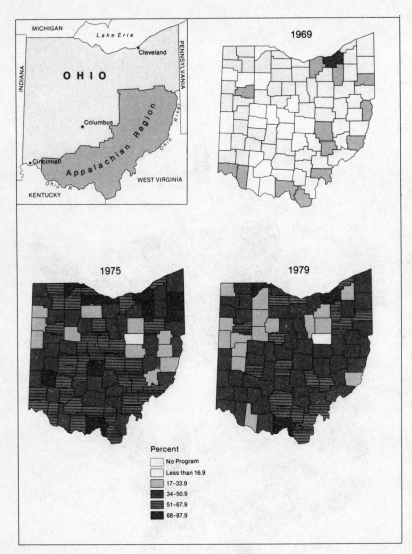

Figure. 8.3. Ohio participation rates

remained at approximately that level in 1979. The mean participation rate was 10.2 percent in 1969, 50.8 percent in 1975, and 43.8 percent in 1979.

FSP effectiveness is accounted for by a stepwise, multiple regression model calibrated separately for 1969, 1975, and 1979, with independent variables pertaining either to underlying demand for food stamps or to the

Program's availability. Thus, our analyses provide insight into the relative role of each type of variable and shifts in that role over the Program's evolution.

Specifically, the model is:

$$PR_{it} = b_{0t} + b_{1t}UR_{it} + b_{2t}YR_i + b_{3t}TEST_i + b_{4t}APP_i$$
$$+ b_{5t}BR_{it} + b_{6t}BL_{it} + b_{7t}URB_{it} + b_{8t}AGE_{it}$$
$$+ b_{9t}REL_{it} + b_{10t}EDUC_{it} + b_{11t}VOTE_i$$

where independent variables are defined as follows:

UR_{it} is the unemployment rate in county i at time t. Variations in need for the program should be accounted for by the dependent variable, since participation rates are expressed in terms of the eligible population. The unemployment rate is included in the study to ascertain whether program use is disproportionately higher in areas of greatest need.

YR_i represents the year in which food stamps became available in county i. The relationship of this variable with PR_{it} is expected to be negative, such that the most recent initiations (highest values) exhibit low program use, common in the early stages of a program's introduction.

$TEST_i$ is a dummy variable, coded 1 if county i was an experimental pilot project, 0 otherwise. Counties which served as pilot projects prior to national policy initiation in 1964 are apt to have high participation rates, since they were chosen on the basis of substantial need, adequate welfare bureaucracies, or high potential for effectiveness (Segal, 1970).

APP_i is also a dummy variable, coded 1 if county i is located with Appalachia, 0 otherwise. Given the prevalence of poverty in this area, participation rates are apt to be high.

BR_{it} is the ratio in average benefit levels between the FSP and another major welfare scheme, the Aid to Families with Dependent Children Program (AFDC), in county i at time t. Because benefit levels for the FSP are set by the national government, while those for AFDC are determined at subnational levels, the ratio is used to examine intergovernmental differences in welfare provision. Specifically, FSP participation rates are expected to be high in those locales where the ratio is high, that is, where food stamp benefits are large, relative to the alternative program.

BL_{it} is the percentage of the population which is black in county i at time t. This variable is included to examine whether participation rates are higher or lower in counties with substantial minority populations.

URB_{it} is the percentage of the county's population living in urban areas. The FSP originally was regarded as an urban-biased program although large numbers of rural residents began to participate throughout the country after 1979 legislative revisions.

AGE_{it} represents the percentage of a county's population living below the poverty level which is older than 65 years. Based upon previous studies (MacDonald, 1977; USDA, 1980), we expect participation rates to be relatively low in counties with substantial poor, elderly populations.

REL_{it} is the percentage of a county's population which identifies with fundamental Protestant religions.[6] Several analyses have demonstrated that welfare use is low in areas where these groups predominate (Hutcheson and Taylor 1973; Stonecash and Hayes, 1981).

$EDUC_{it}$ is the median years of school completed by adult residents of county i at time t. Welfare use tends to be positively correlated with education levels.

$VOTE_i$ is the proportion of a county's voters who supported George McGovern, as opposed to Richard Nixon, in the presidential election of 1972. The variable is used as a surrogate for political liberalism. Although a vote for Nixon did not necessarily suggest conservative political attitudes, a vote for McGovern most certainly indicated a liberal stance on political issues, given the candidate's platform. It is anticipated that FSP use is higher in counties where liberal attitudes predominate.

Results of the analyses are most clearly summarized visually. Figure 8.4 portrays standardized regression coefficients of the independent variables for each year. (Detailed results are presented in the Statistical Appendix.) The changing influence of each factor upon program use will be described in turn.

The unemployment variable (UR) is positive and significant in all three periods, with increases in the magnitude of standardized coefficients through time. Thus, as hypothesized, there is greater than expected program use in counties with highest need, especially in the most recent period. This aggregation effect may be due to more vigorous welfare provision or to decreased stigma in areas of especially high unemployment.

The year in which the program was established in a county (YR) is the single most important factor in 1969. The coefficient is negative and significant; thus, the counties with the oldest projects had the highest participation rates, as expected. However, the importance of establishment date diminishes as the program matures, as seen by the insignificant coefficients in 1975 and 1979. This shift in importance may be explained by the fact that Ohio law had ordered its counties to provide food stamps by 1970, much earlier than the 1974 deadline mandated by national legislation.

The test project factor (TEST) is also significant in 1969, but insignificant

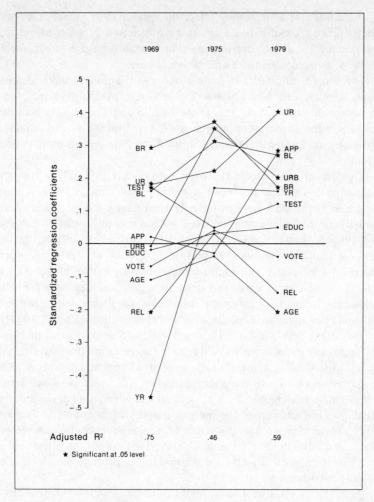

Figure 8.4 Sources of variable program use

in 1975 and 1979. Thus, those counties which were selected as experimental projects prior to the 1964 initiation of the program had higher participation rates early in its development, but the advantage of being a test project has diminished through time.

The Appalachian variable (APP) is insignificant in both of the early periods but positive and significant by 1979. There were several coal strikes in the region in the late 1970s and although strikers comprised only a miniscule proportion of FSP participants, the temporary interruptions in the mines had a depressing effect throughout the regional economy. These difficult conditions are apt to have increased program use in the area, indicating that extensive depression has an aggregate effect, much as unemployment conditions were shown to have.

The benefit ratio (BR) is positive and significant in all three periods. Thus, counties with high FSP benefits, relative to AFDC payments, tend to have high participation rates in the Food Stamp Program. In counties where benefits of the other program are low, individuals in need of public assistance are apparently attracted to the Food Stamp Program in greater numbers.

The urbanization factor (URB) is insignificant in 1969, is strongly positive in 1975, then declines but remains significant in 1979. Welfare programs in the USA have historically been biased toward urban areas. Underlying this relationship between welfare use and urbanism may be such factors as the higher degree of access to welfare offices, due to agglomeration of the poor in particular sections of American cities; the relative efficiency and sophistication of urban welfare bureaucracies, due to economies of scale; and more favorable attitudes toward public assistance in metropolitan areas, where individuals are less dependent on family ties and more dependent on formal institutions (Kodras, 1984).

This urban bias is most evident in the Food Stamp Program in 1975. Its decline by 1979 is apt to be due to a change in the provisions of the policy. Specifically, a major revision of the Food Stamp Act in 1977 made several changes to the policy, including reductions in income levels and deductions allowable for participation and elimination of the requirement that individuals pay for the stamp coupons, beginning in January 1979. These changes had the effect of bringing substantial numbers of rural residents into the program (USDA, 1981). Thus, an alteration in provision factors has affected the distribution of program use, decreasing urban–rural disparities.

Participation rates tend to be higher in counties with substantial black populations (BL) in 1975 and 1979, but the coefficient is insignificant in 1969. Because blacks in Ohio are clustered in the large cities, we might expect the relationship between PR and BL in the latter periods to be a function of their urban location, but the relationship controls for urban effects. The findings indicate program response to an economically disadvantaged group in the recent periods. That the relationship is insignificant in the earliest period may reflect the fact that political consciousness of minority issues, as represented by the overall War on Poverty, had not yet exhibited substantial effect.

The elderly poor factor (AGE) is significant only in 1979. That counties with large proportions of their poor who are older than 65 years tend to show low program use in this latest period is to be expected since, in general, the elderly regard any type of assistance, even that provided by church organizations, as charity. Why this relationship should appear only in 1979 is more problematic. Elimination of the purchase requirement reportedly brought large numbers of the elderly into the Program nationwide (USDA, 1981). Prior to this policy change, many of the elderly hesitated to commit personal resources for the stamp coupons, wanting to reserve their monthly budget for possible medical expenses. Elimination of the purchase requirement apparently did not have this effect in Ohio in 1979.

The proportion of a county's population which is affiliated with fundamentalist Protestant religions (REL) is negative and significant in 1969 but becomes unimportant after that time. Thus, early in the history of the program, participation rates were low in counties with substantial fundamentalist groups. This finding is to be expected because, as Stonecash and Hayes (1981, p. 687) state: 'The impact of fundamentalism appears to stem from the Protestant Puritan heritage which emphasizes industry and productivity and produces very strong feelings on policies relevant to controlling and assisting individuals.' The relationship between PR and REL fades through time, however.

The education level variable (EDUC) is never significant, perhaps because the very small differences in educational attainment at the county level in Ohio were not able satisfactorily to capture variations in the amount of information about welfare which is available to individuals. Finally, the liberal voter factor (VOTE) is never significant. Its zero-order correlations with the dependent variable indicate that participation rates tend to be higher in counties with a liberal political stance, as measured by this variable, but the effect is removed by other factors in the analysis.

4. Concluding observations

Having reviewed the findings for individual factors, we summarize their collective meaning. Although the welfare literature contains few explicit references to diffusion theory, the participation determinants traditionally studied are often interpreted as indicators of demand, for which the *adoption perspective* might serve as a conceptual foundation, or as indicators of supply and availability, for which the *market and infrastructure perspective* would be an appropriate framework. For example, variables such as the unemployment rate or the presence of cultural subgroups typically are considered measures of the need and/or social–psychological disposition toward welfare programs, while variables such as the benefit level or administrative restrictiveness are interpreted as indicators of program provision and availability.

Considered in this context, our analyses suggest that local variations in FSP participation were due primarily to supply-side diffusion forces early in the history of the program, but as these constraints were loosened, demand-side factors have become more important recently. Specifically, the provision factors (benefit ratios, the year in which a county instituted the program, and the existence of early experimental projects) were relatively more important in 1969. By 1979, participation was higher in areas and for groups in greatest need (counties with high unemployment rates, those containing substantial black populations, and those located in Appalachia).

The diminishing effect of provision variables through time, and concurrent ascendance of demand factors, should not be interpreted as evidence that the government eventually loses control of the determinants of welfare use. It is our contention that many of the factors traditionally interpreted as indicators of demand are, in fact, driven by supply. As discussed above in detail, the increasing response of program use to unemployment through time could be due to changes in underlying availability characteristics, such as more vigorous welfare provision in areas of particular need, or to shifts in demand, including a decline in stigmas as an agglomeration of unemployed begin to feel they have a right to government assistance. Even this shift in demand may be influenced by availability conditions if, for example, the decreased stigmas result from a more strenuous advertizing program, and more liberal attitudes among welfare administrators.

In summary, supply forces controlling innovation availability, although they may not be directly measurable in the form of quantitative variables, do not cease to be important determinants of welfare participation as a program matures. Instead, they continue to operate in a lesser role with the demand factors to determine aggregate participation levels. The statistical analyses, like the qualitative account, indicate the importance of program accessibility and availability upon welfare participation and, thus, the *market and infrastructure perspective*, as well as the more traditional *adoption perspective*, is an appropriate conceptual foundation in the examination of welfare program diffusions. The spatial and temporal dimensions of diffusion theory, and its attention to causal forces of provision and demand, place a political innovation in historical and areal context, allowing us to identify the situation in which the program developed and its resultant impact upon regional economies.

With regard to this final point, welfare programs are traditionally regarded as cushioning mechanisms in difficult economic situations, serving areas in particular need for assistance and dampening, to some extent, the effect of regional cycling. On the other hand, this study and a related set of analyses (Kodras, 1982; Jones, 1984; Jones and Kodras, 1984; Kodras, 1984) have shown that the distribution of need for assistance often does not match the pattern of welfare use, indicating that areas receive benefits from the welfare system which are disproportionate to their requirements and suggesting that the role of public assistance in moderating regional inequities is diminished.

The factors which underlie disparities in policy innovation effectiveness (such as variations in political clout, degree of urbanization, and population composition) may also influence the diffusion of technological innovations and their consequent impacts (Malecki, 1983). The idea that these societal forces may have parallel and cumulative effects upon different types of innovation diffusions, and their impact upon regional economies, has not been explicitly examined but constitutes a provocative research issue.

Notes

1. For other studies in this genre, see Agnew, Brown, and Herr (1978); Bingham (1976); Brown and Philliber (1977); Brown, Williams, Youngmann, Holmes, and Walby (1974); Feller, Menzel, and Kozak (1976); and Perry and Kramer (1978). Supply effects upon program adoption have rarely been noted in the literature, although Gregg (1974), Jones (1984), and Kodras (1982) are exceptions.
2. In this regard, Brown (1981, Ch. 3) identifies three organizational modes within which diffusion agencies operate: a centralized decision-making structure; a decentralized decision-making structure; and a decentralized structure with a coordinating propagator. As noted, most entitlement programs fall into this last category.
3. More generally, Brown (1981, Ch. 4) notes four elements of diffusion strategies, all of which have corollaries in public sector diffusion: the development of infrastructure and organizational capabilities, pricing, promotional communications, and market selection/segmentation. The orchestration of these into a diffusion agency strategy depends upon innovation characteristics, agency characteristics, innovation life cycle, and the spatial extent of diffusion.
4. The coefficient of variability is calculated by dividing a variable's standard deviation by its mean. While a standard deviation measures absolute variability, the coefficient indicates relative variability, and thus is useful when comparing different distributions. A large value suggests greater variation among observations in the distribution, a small value indicates lesser variation.
5. The number of eligible persons is measured as the number of individuals living below 125 percent of the poverty level. This surrogate is necessary because the government does not collect data on the actual number of eligibles. See Allen (1977), MacDonald (1977), USDA (1980), and Kodras (1982) for justifications and additional uses of the measure.

6. Data are obtained from the US Bureau of the Census (1973, 1983), the US Department of Agriculture (monthly), and the National Council of Churches (1957).

References

Agnew, J. A., Brown, L. A., and Herr, J. P. (1978). 'The Community Innovation Process: A Conceptualization and Empirical Analysis,' *Urban Affairs Quarterly*, **14**, 3–30.

Allen, J. T. (1977). 'The Food Stamp Program: Its History and Reform,' *Public Welfare*, **35**, 33–41.

Amidei, N. (1981). 'Food Stamps: The Irony of Success,' *Public Welfare*, **39**, 15–21.

Bingham, R. D. (1976). *The Adoption of Innovation by Local Government* (Lexington, Mass., Lexington Books).

Brown, L. A. (1981). *Innovation Diffusion: A New Perspective* (New York, Methuen).

Brown, L. A., Malecki, E. J., Gross, S. R., Shrestha, M. N., and Semple, R. K. (1974). 'The Diffusion of Cable Television in Ohio: A Case Study of Diffusion Agency Location Processes of the Polynuclear Type,' *Economic Geography*, **50**, 285–99.

Brown, L. A. and Philliber, S. G. (1977). 'The Diffusion of a Population-Related Innovation: The Planned Parenthood Affiliate,' *Social Science Quarterly*, **58**, 215–28.

Brown, L. A., Williams, F. B., Youngmann, C. E., Holmes, J., and Walby, K. (1974). 'The Location of Urban Population Service Facilities: A Strategy and its Application,' *Social Science Quarterly*, **54**, 784–99.

Citizens' Board of Inquiry into Hunger and Malnutrition in the United States (1968). *Hunger, USA* (Boston, Beacon Press).

Coe, R. (1980). 'A Preliminary Empirical Examination of the Dynamics of Welfare Use,' *Five Thousand American Families: Patterns of Economic Progress* (Ann Arbor, Michigan, The Institute for Social Research, The University of Michigan).

Collier, D. and Messick, R. E. (1975). 'Prerequisites versus Diffusion: Testing Alternative Explanations of Social Security Adoption,' *American Political Science Review*, **69**, 1299–1315.

Feller, I. and Menzel, D. C. with Kozak, L. A. (1976). *Diffusion of Innovations in Municipal Governments* (University Park: The Pennsylvania State University, Center for the Study of Science Policy).

Foster, G. M. (1962). *Traditional Cultures and the Impact of Technological Change* (New York, Harper).

Gray, V. (1973). 'Innovation in the States: A Diffusion Study,' *American Political Science Review*, **67**, 1,174–85.

Gregg, P. M. (1974). 'Units and Levels of Analysis: A Problem for Policy Analysis in Federal Systems,' *Publius: The Journal of Federalism* 4:4, 59–86.

Hutcheson, J. D. and Taylor, G. A. (1973). 'Religious Variables, Political System Characteristics, and Policy Outputs in the American States,' *American Journal of Political Science* **17**, 414–21.

James, D. B. (1972). *Poverty, Politics, and Change* (Englewood Cliffs, New Jersey, Prentice-Hall).

Jones III, J. P. (1984). 'A Spatially-Varying Parameter Model of AFDC Participation: Empirical Analysis using the Expansion Method,' *The Professional Geographer,* 36:4, in press.

Jones III, J. P. and Kodras, J. E. (1984). 'AFDC Participation Dynamics and Policies,' *Modeling and Simulation,* 14, in press.

Kodras, J. E. (1982). *The Geographic Perspective in Social Policy Evaluation: A Conceptual Approach with Application to the U.S. Food Stamp Program,* Ph.D thesis, Department of Geography, The Ohio State University, Columbus, Ohio.

Kodras, J. E. (1984). 'Regional Variation in the Determinants of Food Stamp Program Participation,' *Environment and Planning,* 2, 67–78.

Lucas, A. (1982). 'Integrating Analytic Paradigms for Public Policy Diffusion Research,' unpublished research.

MacDonald, M. (1977). *Food, Stamps, and Income Maintenance* (Madison, Wisconsin, Institute for Research on Poverty).

Malecki, E. J. (1983). 'Technology and Regional Development: A Survey,' *International Regional Science Review,* 8:2, 89–125.

National Council of Churches (1957). *Churches and Church Membership in the United States* (New York: NCC).

Perry, J. L. and Kramer, K. L. (1978). *Diffusion and Adoption of Computer Applications Software in Local Governments* (Irvine, University of California at Irvine, Public Policy Research Organization).

Piven, F. and Cloward, R. A. (1971). *Regulating the Poor: The Functions of Public Welfare* (New York, Random House).

Rogers, E. M. (1969). *Modernization Among Peasants: The Impact of Communication,* (New York, Holt, Rinehart, and Winston).

Rogers, E. M. and Shoemaker, F. F. (1971). *Communication of Innovations: A Cross Cultural Approach* (New York: Free Press).

Ross, M. H. and Homer, E. (1976). 'Galton's Problem in Cross-National Research,' *World Politics,* 29, (October), 1–28.

Segal, J. A. (1970). *Food for the Hungry: The Reluctant Society* (Baltimore, Md, The John Hopkins Press).

Stonecash, J. and Hayes, S. W. (1981). 'The Sources of Public Policy: Welfare Policy in the American States,' *Policy Studies Journal,* 681–98.

Thorkelson, H. (1969). 'Federal Food Programs and Hunger,' in J. Larner and I. Howe (eds), *Poverty: Views from the Left* (New York, William Morrow), 184–95.

US Bureau of the Census (1973). *1970 Census of the Population* Washington, DC, US Government Printing Office).

— (1983). *1980 Census of Population* (Washington, DC, US Government Printing Office).

US Department of Agriculture. *Statistical Summary* (Washington, DC, US Government Printing Office), monthly.

— (1980). *The Food Stamp Progam: Income Maintenance or Food Supplement?* (Washington, DC, US Government Printing Office).

— (1981). *Effects of the 1977 Food Stamp Act: Second Annual Report to the Congress* (Washington, DC, US Government Printing Office).

US Senate, Ninety-first Congress, first session. Select Committee on
    Nutrition and Human Needs (1969). 'Poverty, Malnutrition, and
    Federal Food Assistance Programs: A Statistical Summary,' (Wash-
    ington, DC, US Government Printing Office).
Walker, J. (1969). 'The Diffusion of Innovations among the American
    States,' *American Political Science Review, 63,* 880–99.
West, G. (1981). *The National Welfare Rights Movement,* (New York,
    Praeger).
Wohlenberg, E. H. (1976a). 'Interstate Variations in AFDC Programs,'
    Economic Geography, 52:3, 254–66.
— (1976b). 'Public Assistance Effectiveness by States,' *Annals, Associa-
    tion of American Geographers,* 66:3, 440–50.

Statistical appendix

	1969		1975		1979	
Independent variables	β	r	β	r	β	r
UR_{it}	0.18*	0.14	0.22*	0.07	0.40*	0.51
YR_i	−0.47*	−0.76	0.17	0.21	0.16	−0.25
$TEST_i$	0.17*	0.46	0.05	0.31	0.12	0.35
APP_i	0.02	0.12	−0.03	−0.22	0.28*	0.10
BR_{it}	0.29*	0.55	0.37*	0.52	0.17*	0.43
BL_{it}	0.16	0.56	0.31*	0.57	0.28*	0.53
URB_{it}	−0.01	0.45	0.35*	0.53	0.20*	0.40
AGE_{it}	−0.11	−0.37	−0.04	−0.28	−0.21*	−0.30
REL_{it}	−0.21*	−0.27	0.03	−0.19	−0.15	−0.35
$EDUC_{it}$	−0.02	0.01	0.03	0.11	0.05	0.16
$VOTE_i$	−0.07	0.44	0.04	0.35	−0.04	0.41
Adjusted R^2	0.75		0.46		0.59	

* Indicates that the variable is significant at the 0.05 level or better.

9 The Impact of New Information Technology on Urban and Regional Structure in Europe*

J. B. GODDARD, A. E. GILLESPIE,
J. F. ROBINSON AND A. T. THWAITES
University of Newcastle-upon-Tyne

1. Introduction

The impact of developments in communications on urbanized regional economies has been the subject of much debate, but in the main the principal emphasis of this discussion has been placed upon physical communication—the movement of goods and people—rather than the movement of information. Technical advances in transportation have clearly had a profound effect on cities and regions. In the nineteenth century railways made possible the development of industries adjacent to mineral resources and residential areas expanded nearby, leading to urban agglomerations. Subsequent improvements in road communications during the twentieth century facilitated a dispersal of industry from cities and permitted an increasing separation of homes and work places. Similarly, new handling methods such as containerization undermined the entrepôt functions of city-centre rail and port facilities. So in a very general sense subsequent advances in physical communication have reduced the earlier rationale for urban concentrations within regions.

But what of the broader aspects of communication that relate to the flow of information between organizations and individuals? Far less is known about the impact of an increasing ability to move large amounts of information from one location to another on the future development of the city, and the wider urbanized regional economy. Meier some time ago articulated his 'communication theory of urban growth', while Webber has suggested that improved possibilities for communication have created a 'non-place urban realm' (Meier, 1962; Webber, 1964). However, few historical studies have been made of the impact on cities of improvements in information transfer made possible by the conventional handset telephone. The evidence that is available suggests that the telephone may have been a powerful centralizing force because it facilitated the growth of large companies, which in turn located headquarters functions in the

* This paper draws heavily on a study undertaken at the Centre for Urban and Regional Development Studies, University of Newcastle upon Tyne on the 'Effects of New Information Technology in the Less Favoured Regions of the Community' for DGXVI of the European Commission.

centres of the major cities of regional economies (e.g. Pool, 1978, Prais, 1976). Thus the telephone made it possible for large enterprises to control productive activities in a variety of regional locations from a central office; at the same time it also assisted communications within the city-centre skyscraper. The telephone may therefore have been an important factor in urban and regional industrial concentration during the 1930s and the consequent development of central business districts.

The most obvious question for this final chapter is whether more recent advances in information technology, particularly the convergence of telecommunication and computing, will break down this centralizing effect which has ensured the survival of cities and the core regions of national economies as office centres. But in addition, there are a number of more subtle and wide-ranging issues relevant to urban and regional development that are raised by advances in information technology. In order to understand these broader matters it is necessary to understand the nature of the core element in information technology, the micro-processor, and its revolutionary nature which widens the 'potential impact' question far beyond the simple improvement of the telephone system.

As will be argued, the word 'revolutionary' is appropriate in its historical context, for the microprocessor is producing an economic impact similar in importance to the steam engine, which played a central role in the first industrial revolution. As with the steam engine, this technological impact is arising through the widespread diffusion of the new technology (Freeman et al., 1982). Just as the steam engine moved from a limited application, such as pumping water, to becoming incorporated into a wide range of applications in transport and manufacturing, so too has the computer moved out of its air-conditioned 'temple' into applications in a wide range of products and services.

In studying the impact of technical change on economic development Schumpeter made an important distinction between product, process and managerial innovations (Schumpeter, 1939). The revolutionary nature of the microprocessor arises from the fact that it is influencing all three of these dimensions of technological change simultaneously. As a product innovation, the microprocessor is already affecting *what* goods and services are provided, and as a process innovation it affects *how* these goods and services are provided and delivered. As a managerial innovation, the microprocessor is influencing the way in which organizations are structured and controlled. These changes feed through into the structure of employment, influencing the industries in which people work (i.e. product classification of employment) and their occupations (process classification of employment). Last but not least, the changes will directly and indirectly affect *where* jobs are located.

This perspective suggests that employment changes will arise through a process of job displacement in certain sectors and occupations, and job creation in other sectors and occupations. These changes will occur in particular places with job creation and job displacement being indissolubly linked. For example, in localities where industries fail to introduce new processes and products, jobs will be lost through differential growth to industries in competing areas which are able to produce existing products at lower prices or provide more attractive new products to perform identical functions. Such sectorally-based job losses may be greater than those resulting from the *direct* displacement effects on particular occupations of introducing new process technology. A few jobs may be directly *relocated* between areas as a result of managerial innovations which facilitate new forms of corporate organization. Similarly, process innovations may lead to new skill requirements that can be most economically provided in alternative areas. However, in this process fewer jobs may be created in the new location than are displaced in the old. Similarly, new services (e.g. computerized information services for business management) may displace conventional service jobs, but may also create service employment through the realization of previously latent needs.

But what is the *time-scale* over which such processes might operate? Some jobs may be displaced in one period as a result of productivity gain; this may indirectly enable new jobs to be created at another time *and* in another place. Such changes may occur as a result of the location decisions of individual companies, or may 'work themselves through' the economy as a result of macroeconomic effects on changing patterns of consumer demand. Moreover, what is the *spatial scale* over which such processes might operate? Most changes will occur within national economies, but increasingly may involve new employment locations in a number of countries. For example, standardization of production may enable jobs to be 'exported' to other nations. At the same time, further structural changes will occur on a city scale, and may even differentiate inner city locations from the suburbs.

These questions suggest that, while the future of particular occupations or industries can be to some degree anticipated, the interaction of occupational, sectoral and locational effects is even more difficult to ascertain. In the light of such difficulties, the Commission of the European Community, in attempting to assess the impact of new information technology (NIT) on employment, has suggested:

The absolute job creation potential of a technology can never be measured even with hindsight and *a fortiori* when it so profoundly affects economic activities. It is therefore more useful to concentrate upon the adaptive efforts necessary within the whole economy to realise the potential of the technology and thus promote full employment in Europe (EEC, 1982).

The critical point for the concerns of this chapter is that the success of this adaptive effort will be deeply conditioned by the infrastructure, industrial and institutional capacity of individual cities and regions. Areas which lack the necessary telecommunications infrastructure, innovative enterprises and public and private agencies will lag behind in the race to take advantage of the employment opportunities offered by information technology. The remainder of the chapter discusses some key elements which could influence the adaptive capacity of regions and their constituent cities, starting with the necessary modernization of the telecommunications infrastructure, and then moving on to consider the impact of information technology on the location of industrial and service activities.

2. Telecommunications networks

Most observers regard telecommunications networks as the 'highways' of the future. As with roads, such networks are a permissive factor in economic development—that is a necessary but not sufficient condition; any shortcomings in the telecommunications infrastructure of a particular city relative to other cities will inhibit its development. But having these facilities will not ensure that development takes place if there is not the sufficient entrepreneurial capacity to respond to the opportunities provided by that infrastructure. Since telecommunications are inherently an international activity, differences between countries in the quality of their telecommunications network will also be a major factor in the development of a country's constituent cities and regions. So the possibilities for development in particular regions must be seen in their national and international context.

The major developments that are occurring in telecommunications can be divided into networks, switching and services, but in each respect the most important transformation is from analogue to digital communications. Eventually, most countries will possess an Integrated Services Digital Network (ISDN) with distance-independent tariffs and computer-controlled switching. In other words, the telephone will, in effect, become a computer terminal, giving access to a worldwide computer network. When such a network is in existence the forces for agglomeration will be limited. However, the stages in which ISDN, and the services which it makes possible, are introduced are likely to favour the existing concentrations of population.

2.1 *Transmission networks*

As far as transmission is concerned, the most dramatic reduction in costs has occurred in long-distance communications. The recent introduction of fibre optics cable has further reduced these costs. However, with low utilization factors the cost per user, or potential user, of installing fibre-optics cabling in non-urban areas is extremely high. This has meant that the main initial benefits of this technological advance have occurred in inter-city as opposed to intra-city communications. While the higher density inner-city areas are generally cheaper when it comes to installing new cable networks, inner-city residents are often relatively poor and the proportion of users likely to subscribe to new networks may be below that in lower density suburbs. While these developments favour cities, the actual impact will depend on pricing policies and the extent of cross-subsidization of local communications by trunk traffic. Deregulation of telecommunications in the United States has reduced this cross-subsidization and this may paradoxically favour urbanized regions at the expense of rural areas (Langdale, 1983). Deregulation of networks is also occurring to a limited extent in the United Kingdom with the licensing of a competitor to British Telecom. Project Mercury will provide a fibre-optic, competitive network linking a number of major cities; other cities outside this routeway will continue to be faced with a monopoly provider of communication services (Fig. 9.1).

It is easy to exaggerate the scale of the immediate impacts of network modernization. At the present time improvements in trunk transmission are not directly impinging on the majority of users, simply because conventional copper cables at the extremities of the network mean that transmission speeds are reduced by the analogue/digital interfaces. Only in a number of experiments, such as BIGFON in Germany, are digital 'end-to-end' services using broad-band optical fibre cables provided for selected users. Because of the need to modernize the entire network step-by-step, most telecommunications administrations in Europe have introduced overlay networks which provide high speed and low cost transmission, but between a restricted number of locations. Such networks are entirely digital and designed to carry data with charging based on volume rather than distance. Examples of such systems are Packet

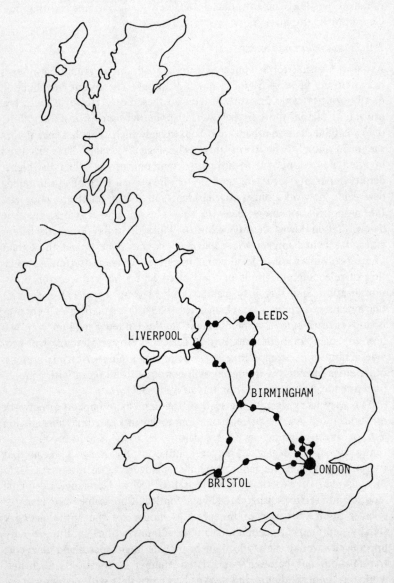

Figure 9.1 The United Kingdom's proposed Mercury network. *Source*: CS & P London.

Switching Service (PSS) in the United Kingdom (Fig. 9.2) and TRANSPAC in France. The important point for urban and regional development is that such networks generally serve a restricted number of locations, usually the largest cities. Users outside these cities are forced to make do with either the high cost and low speed analogue Public Switched Telephone Network (PSTN) or to lease private lines from the PTT.

It has been suggested that satellite transmission will overcome these problems of sequential network modernization. However, this option is only feasible for the very large users when it comes to transmission as opposed to simply receiving. There are nevertheless important implications for individual cities arising from global satellite communication through the impact of this technology on individual multinational corporations. In many sectors, such as the motor vehicle industry, internationally organized production is dependent on cheap data communications; cities which are fortunate enough to have the headquarters operation of such companies can clearly benefit from developments in global communications. On the other hand, cities with regional offices may find themselves bypassed as companies now deal directly with production plants which may be located outside the cities. More generally, the development of specialized data networks and the use of leased lines which are preceding the full introduction of ISDN result in benefits of the reduced cost of telecommunications accruing to the largest business users and not the small to medium-sized enterprises which tend to characterize the economy of certain large cities.

2.2 *Advances in switching*

The keynote in switching is the introduction of fully electronic exchanges to replace the electro-mechanical 'step-by-step' Strowger exchanges. Such exchanges are essentially computers and the important point for the end-user is that they make possible a wide range of enhanced services such as automatic transfer of calls. In terms of urban and regional development, it is important to appreciate that the modernization of switching as well as networks will occur on an incremental basis. Areas that have in the recent past received investment in intermediate semi-electronic exchanges will have to wait until this equipment is amortized before receiving the fully electronic facility. Those areas with the new capacity will then be able to provide new services and thereby gain short-term comparative advantages. Although the time lags in the progress of exchange modernization may be limited, these lags provide important windows of opportunity for enterprises located in the favoured areas.

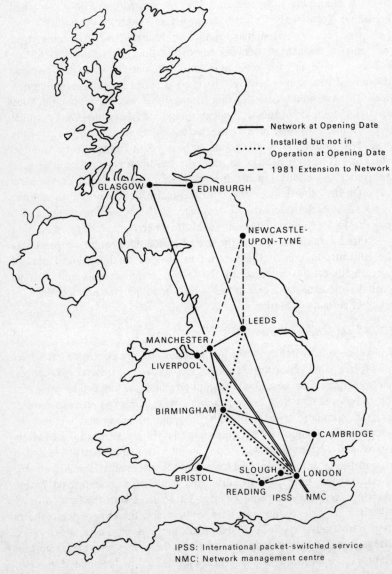

Figure 9.2 The United Kingdom's Packet-Switched Service (PSS). *Source*: POEJJ.

2.3 New information technology-based telecommunications services

From the point of view of the user, modernization of networks and switching is only important in so far as it influences the availability and costs of telecommunications-based services. The historical experience has been that new services are introduced by PTTs in the largest cities and these services are subsequently diffused down the urban hierarchy (Clark, 1978; Langdale 1983). A more recent service, Videotex (or Prestel in the UK), which is intended to be a distance-independent service, illustrates a similar point; because the coverage of access points is not sufficiently dense to make possible universal local call charging, areas outside the largest cities are at a disadvantage.

In Britain and Italy the provision of services like Videotex is no longer a PTT monopoly; so-called 'value added networks services' (e.g. message services) can be provided by private businesses to third parties. This is generally regarded as a significant area of job generation related to new information technology, but depends to some degree on the necessary modernization of the network. Regions with the appropriate infrastructure may be able to develop these services based on local demand and, as links to other areas are introduced, displace more traditional services in these localities.

Taking all these points together, there is little to indicate that development in telecommunications networks is likely to disadvantage the largest cities relative to small towns and rural areas in any national urban and regional system. The incremental modernization of networks and the logic of density will ensure that the inner parts of large cities will have an initial advantage. However, it will be necessary for enterprises in such areas to respond to opportunities offered by new networks and services. Extensive evidence on the take-up of traditional telephony, telex and data terminals attached to the PSTN indicate that economically lagging areas of Europe have the lowest penetration levels (Fig. 9.3). While convergence between areas is occurring in the uptake of traditional services, a new round of divergence is appearing in relation to new developments. These in part stem from a lower level of awareness of the opportunities offered by new information technology. For example, Table 9.1 shows a lower level of awareness of the United Kingdom's Videotex system in the least favoured areas such as the North East of England than in the core regions located in the South of the country. In an increasingly commercial environment, PTTs will not continue to provide services in advance of demand if that demand fails to come forward; lagging regions will therefore find themselves at a greater disadvantage.

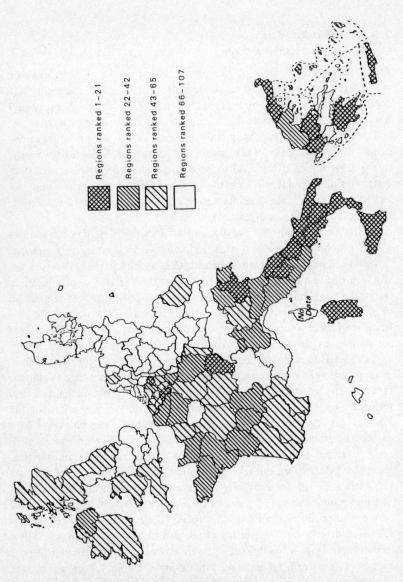

Figure 9.3 Index of telecommunications penetration*

  * Derivation of the index: each of the 107 regions are first ranked in ascending order in terms both of their telephone and telex equipment penetration levels (standardized by population). The two rankings are added together, and the composite index so produced is in turn ranked in ascending order from 1 to 107. This ranking is mapped above, such that the heaviest shading (regions ranked 1–21 on the index) are those with the lowest levels of telecommunication equipment penetration.

Table 9.1 Regional levels of awareness of PRESTEL (the
UK's Videotex service)
in the first quarter of
1982

London	44%
East Anglia	41%
South East	39%
West Midlands	33%
South West	32%
Yorkshire	25%
East Midlands	23%
North West	22%
Scotland	19%
North	18%
Wales	14%
Northern Ireland	n/a

Source: Prestel.

2.4 *The international dimension to network development*

While the above discussion has highlighted differences between large cities and other locations within a nation, such differences can be less significant than international contrasts. There are marked variations between the countries of Europe in the extent of network modernization and in charging policies. The development of international overlay networks like Euronet has favoured the capital cities (Fig. 9.4). The proportion of telephone exchanges using intermediate technologies such as 'Cross-Bar' also varies significantly between countries, as does the penetration of conventional services like the telephone and telex and advanced services such as Videotex. Last but not least, there are important contrasts in charging for similar services in different countries. Figure 9.5 is based on the *fixed* cost of terminal equipment (telephone, telex and data) and the *variable* costs of making average-length calls to each EEC country for a hypothetical firm with five telephone lines, one telex machine and one data terminal. The variations are considerable, with Luxembourg's cost index being only 43 per cent that of Ireland, the most expensive location. Thus, generally speaking, the peripheral countries of Europe are still disadvantaged in spite of the apparent 'distance shrinking' capacity of telecommunications. Figure 9.6 shows that the highest cost countries, with the exception of Germany, are generally the poorest. Variations within countries are therefore superimposed on these international contrasts.

The significance of these differences arises in terms of their impact

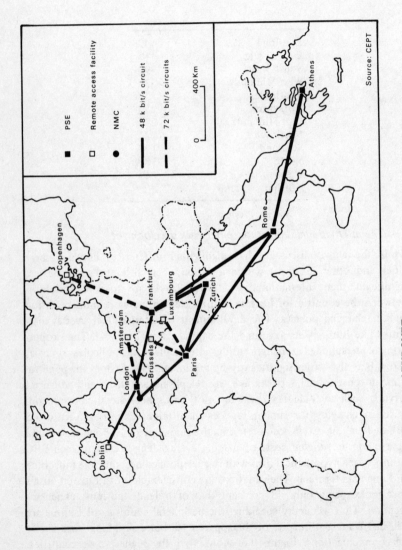

Figure 9.4 The configuration of EURONET

Legend:

- ■ PSE
- □ Remote access facility
- ● NMC
- ── 48 k bit/s circuit
- ┄┄ 72 k bit/s circuits

0 ─── 400 Km

Source: CEPT

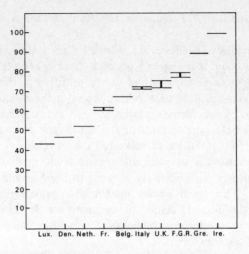

☐ Indicates highest and lowest regional values

Figure 9.5 An index of telecommunication cost variations between the member states of the European Community.* *Source*: CURDS Final Report to the EEC on the 'Effects of New Information Technology on the Less Favoured Regions of the Community', 1983.

* The index is expressed in *relative* terms such that the most expensive region (Ireland) is given the value of 100. It should be noted that the index measures the 'average' cost of making calls from each region to each other region in the Community, weighted by their populations irrespective of the distances involved. The results are consequently illustrative of tariff differences rather than representative of expenditure under actual traffic patterns.

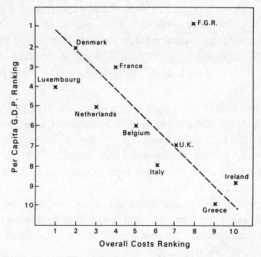

Figure 9.6 Relationship between the telecommunication cost index for member states and their per capita Gross Domestic Products. *Source*: As for Figure 9.5.

on the profitability of enterprises. A small firm in the commercial and service sector could spend up to 5 per cent of turnover on telecommunications equivalent to perhaps a third of net profits. For the hypothetical Euro-firm making calls to each other country in the community, a location in Ireland would depress profits by -13 per cent, while a location in Denmark would increase profits by $+12$ per cent. In addition to these immediate effects, a failure to make full use of telecommunications for either cost, availability or awareness reasons could mean that the enterprise concerned is not effectively scanning its business and technological environment. This could inhibit market and technological innovation and lead to the ultimate demise of the enterprise. It is this question of the response to information technology as a product and process innovation to which we now turn.

3. The impact of new information technology (NIT) on manufacturing industry

3.1 *The manufacture of new information technology products*

There can be no doubting the global importance in terms of economic opportunities of the manufacture of new information technology. The annual world market is increasing at around 14 per cent per annum. However, this business is dominated by a few large American and Japanese companies, with only four European firms appearing in the world's top twenty in this field. Nevertheless, there remain opportunities for new, small high-technology companies that are well represented in some urbanized regional economies of Western Europe. According to the OECD (1981), factors favouring small and medium-sized enterprises (SMEs) in this sector are:

- the reduced role of manufacturing in relation to R & D in new information technology manufacturing, which reduces the relative importance of production scale economies;
- the increased rate of technological innovation, which makes it increasingly difficult for any one large firm to maintain technological leadership for any length of time;
- the redefinition of markets, brought about through the convergence of information transmission and processing technology, which has led to a considerable overlapping of interests of firms in previously distinct market areas;
- a reduction in the optimal scale of production, brought about by a fall in fixed capital requirements.

Set against these factors are perhaps more powerful counterveiling forces favouring industrial concentration in large enterprises. These include:

- the greater inexperience of small and medium-sized enterprises (SMEs), as opposed to large corporations, in handling communications and telecommunication resources;
- oligopolistic reaction to increasing competitive pressure, including more intensive R & D spending, monopolization of sales outlets and absorption of new products from small firms through take-overs;
- monopsonistic situations prevailing in most national telecommunications markets where PTTs control the investment strategies of major firms and effectively exclude SMEs from important sections of NIT markets.

The locational implications of such concentration is that it would be difficult for individual cities or their broader regions already lacking manufacturing firms with expertise in the information technology industry to break into these new markets. On a European scale, there is evidence that a few areas have the necessary concentration of company research centres, skilled labour, subcontractors and social capital. Large firms have developed 'production hierarchies' with the routine production of goods at later stages in the product life cycle dispersed to low labour-cost areas outside the major production regions. In the United Kingdom, for example, now outdated, electro-mechanical exchanges were manufactured in the old industrial city of Sunderland using the metalworking skills developed for a previous generation of heavy engineering industry. The new generation of electronic exchanges will be manufactured using significantly fewer workers in another city, Liverpool, which has a concentration of R & D facilities owned by the company enacting these changes (Peck and Townsend, 1984).

Similar generalizations apply to the software side of the NIT industry with a rapid rate of sectoral growth, a large number of new firm entries, but a rapid trend towards concentration. In employment terms, the industry is not currently very significant—34,000 employees in France, a similar number in the United Kingdom, 18,000 in Germany, and about the same number in Italy. The industry is also highly concentrated geographically because the bulk of demand emanates from corporate headquarters located in a few key cities. Areas dominated by branch factories are unlikely to generate a great deal of demand for computer software services. Nevertheless, the close contact necessary between customers and suppliers suggests that most cities would generate some demand for computer software services.

3.2 *The application of new information technology in the small firm sector*

A further dimension to information technology in manufacturing will be its impact as a managerial innovation on the structure of companies and the implication this has for the location of various information-based activities. In the case of SMEs, opportunities for management innovation will arise through the use of new information technology to improve the firm's access to specialist information of a technical and commercial nature. For individual regions and their constituent cities, the critical issue will be the extent to which small enterprises indigenous to the area incorporate the new technology as a managerial innovation in scanning their business environment.

One question is whether by using information technology the SMEs in a declining industrial area can break out of their local industrial environment and thereby gain access to the international body of information related to technical innovation. Unfortunately, survey evidence tends to suggest that SMEs, particularly in lagging areas, fall behind larger firms in the use of information technology as a managerial innovation (Goddard, 1983). This is partly a matter of cost but also reflects a lack of general orientation towards the purposive scanning of the technical environment. In order to encourage SMEs to take advantage of information technology, research suggests that intermediaries are needed between the firm and the vast array of specialist information available on computerized data bases—for example, on patents (Gibbs and Thwaites *et al.,* 1983). In situations in which the specialist adviser/broker is able to build up a relationship of trust with the small firm, and react in an iterative way to a particular problem, very substantial technological benefits and cost savings can accrue. The lesson of this experience is that the availability of new technology alone is not sufficient; small firms will simply have neither the time nor inclination to use it unless it is marketed by specialist intermediaries who can prove its relevance in working situations.

3.3 *Information technology and the structure of large manufacturing organizations*

Small and medium-sized enterprises in most cities, however, account for a small proportion of the total manufacturing sector. The component parts of large enterprises, be they headquarters, subsidiaries or branch plant establishments, are generally far more significant in the urban and regional industrial economy. There is considerable debate about the possible influences of new information technology on such multi-site enterprises, specifically concerning its effect on the degree of centralization or decentralization of administrative functions. It is tempting to translate this debate into spatial terms, given the tendency for corporate headquarters functions to be centralized in capital cities. Before such a translation is undertaken, however, it is initially helpful to map the corporate hierarchy in urban terms and identify the range of functions present in cities. Such analysis may or may not reveal an isomorphism between centralization in corporate and geographical terms.

The complexity of the relationship between technological change and organizational structure in control can be seen in the following extended passage from OECD (1981):

New information technologies tend to integrate the activities of the production process with those of the organisational process. Previously separate control systems are merged to form an overall (process and management) control system. The implication of such structural applications are far-reaching, and as yet, barely perceived, but a number of case study effects can be indicated:

- Organisation units tend to be more integrated and to merge into large departments.
- Programming functions and units tend to be centralised and go upwards in the hierarchy.
- At lower levels of the organisation, more and more autonomy is given to work groups by decentralisation of certain control decisions. On the other hand, there is a centralisation of control of production data.
- The role and significance of middle management is greatly reduced. Some of its functions are now performed by the information system and are controlled at a higher level; some other functions are given to lower organisational levels. Consequently some middle management positions disappear.

Certainly there is evidence that electronics technology has increased the degree to which tasks are structured, limiting discretion in most of the dimensions examined. . . Some of the impacts may take a long time to manifest themselves. For example, upward occupational mobility might be adversely affected. The traditional route from the shopfloor is via supervisory positions which, as suggested, would be less in number and will increasingly require fewer managerial skills (OECD 1981).

It is, however, important to realize that the technology itself in no sense 'causes' a change in organizational relationships; it *makes possible* different forms of organizational relationship, and can consequently be used by adopting companies to support *either a strategy of centralization or decentralization.*

Case studies of individual companies highlight the need to consider the relative importance of communications *vis-à-vis* other factors in decisions about which functions to locate where in a large manufacturing company. Whether more widespread use of NIT will lead to a decentralization of decision-making functions from headquarters in a particular city depends not only on communications factors. Other factors relevant to locational decisions include wages, rents for office space, and travel costs for meetings that must be conducted 'face-to-face'.

Research on the interrelationship between these factors shows that in an analysis of the trade-off between the extra business travel costs arising from relocation from London against the savings on rents and salaries:

— savings do not increase beyond 100 miles;
— business travel costs continue to increase with distance and thereby consume all other savings; and
— only those functions requiring little contact with the capital can be relocated. Such functions tend to be the lowest grade clerical applications, rather than key functions like R & D, or marketing (Goddard and Pye, 1977).

These relationships go some way towards accounting for the predominant pattern of short distance, intraregional office dispersal that has been experienced from London and other major capitals. Clearly, the greater use of telecommunications could modify the costs and benefits in relocation decisions. However, the volume of business travel replaced by telecommunications would have to be substantial to fundamentally alter the influence of other factors. Fitting the range of substitution values obtained from surveys of business contacts into the trade-off analysis described above suggests that the impact of telecommunications on the decentralization of headquarters functions is likely to be marginal.

This conclusion could be modified if relocation were to be associated with some administrative reorganization which led to the devolution of decision-making responsibility to lower levels in the company. More specifically, greater autonomy at lower levels in the company would reduce the need for communications with headquarters. In addition, if relocation were to a major office centre some distance from the capital, rather than a small suburban site, then the likelihood of new business links being forged in the new location would be increased, thus reducing communication costs.

A move away from centralized data processing with remote job entry terminals to distributed processing based around major nodes, *could* support such a strategy when voice and information transmission systems are merged within the company. However, it must be re-emphasized that the technology is only an enabling factor, and that the impact of its use will be dependent on the wider organizational and locational strategies being pursued by individual corporations.

It is important to realize that the status of corporate functions in cities lower down the urban heirarchy may be raised not only by the actual relocation of functions from headquarters centres, but also by the development of managerial functions *in situ* in subsidiary and branch establishments that might be feasible with a greater use of technology. Thus the use of new telecommunications technology as an alternative to business travel could considerably reduce the costs of operating a communication-intensive function in a subsidiary company in a smaller city, as compared to a capital city. While such possibilities exist, the evidence that is available for the United Kingdom suggests that more widespread use of new telecommunications technology in branches and subsidiaries in less developed regions is providing the opportunity for *more* 'remote control' of production. These conclusions emerge from studies of business travel which suggests a greater potential for telecommunications substitution for face-to-face meetings in the case of travel from headquarters to branch establishments rather than in the reverse direction (Goddard and Marshall, 1983). As a result more services are provided within the corporate headquarters in the core region for manufacturing plants in peripheral regions, thereby reducing demand for business services provided face-to-face in the central city of the peripheral region. A declining demand for business services in peripheral regions may lead to a reduction in supply to the detriment of the SMEs sector which characteristically relies on services provided in the local area. So centralization within the corporate sector facilitated by new information technology may have negative implications for locally controlled industry in peripheral regions.

4. The impact of new information technology on the service sector

4.1 *Consumer services*

The largest consumer service is retailing. This sector, and in particular the largest organizations within it are major consumers of new information technology. For example, new technology is used in the form of automated stock control and point-of-sale terminals. While it is clear that the adoption of such labour-saving technology will proceed rapidly in high labour-cost cities, most large operators also recognize that scale economies arise if *all* stores are linked together as quickly as possible. As a result, any direct long-term differentiation in urban employment effects is unlikely to be significant. However, the indirect employment effects may be more important. In established national retailing chains the use of information technology in centralized purchasing and distribution may reduce the importance of urban middle management functions. At the same time, information technology may enable national chains to enter into regional markets previously protected by the barriers of distance. In both instances, established patterns of local purchasing by regional retail chains may also be disrupted. In so far as information technology facilitates further concentration of ownership in the retail sector, it may facilitate further import penetration, both regionally and nationally, with consequent negative impacts for regional and national producers of consumer goods.

Even more subtle indirect effects may arise through interregional variations in the rate of household adoption of new information technology. Such household adoption may be necessary to realize the economies of scale required to justify investment in network infrastructure. This investment will, in turn, provide the basis for more specialized producer services. At the same time, household use of information technology may facilitate business applications through greater familiarity on the part of employers and employees with the potential of the technology. Given that adoption of new information technology by households is likely to be related to income, and, given the existence of higher per capita incomes in the urban centres of core regions, such developments will contribute a further negative element to the disadvantage of the less developed areas.

A further thesis concerning the impact of new information technology on service provision generally also needs consideration in an urban context. This thesis has been developed by Gershuny (Gershuny and Miles, 1983), who suggests that developments in new information technology will accelerate the growth of the 'self-service economy' whereby households substitute purchased technology for services previously bought in established consumer markets (e.g. television sets substituting for entertainment services). This trend may displace jobs in conventional personal services but lead to additional jobs in the creation and management of the 'software' necessary to deliver these services electronically.

Unfortunately, the types of skills necessary to produce such services are highly specialized and available chiefly in the more prosperous regions. Once the electronic services software is written it is perfectly mobile between cities. Thus we may witness locally provided conventional consumer services in the least favoured areas being displaced by new information technology-based services produced and delivered from more prosperous areas. (The situation is analogous to the displacement of local theatres by TV programmes produced in capital city studios.)

Such trends are not inevitable. Information technology could equally facilitate the development of local electronic services tailored to meet local needs. But this depends on the existence of the necessary networks of interest groups coming together to exploit the potential of the new technology. Unfortunately, in many of the less advanced regions, such networks may have atrophied, with the regions becoming highly dependent on more prosperous areas, local initiatives and entrepreneurship having been drained away by the increasing external control of local economies and an increased reliance on transfer payments. Considerable stimulation by public agencies may be necessary to create the conditions favourable for local communities to respond to the potential of new information technology.

4.2 *Producer services*

As will be apparent from the preceding discussion, the impacts of NIT on producer and consumer services are difficult to disentangle from one another. This is particularly the case in banking which serves both business and individual customers. As far as banking services to households are concerned, tele-banking will undoubtedly reduce the number of jobs in bank branches, jobs which largely followed the distribution of population. At the same time, more specialized and personalized services may be provided by banks in a more limited number of urban centres. In addition, some central administrative responsibilities may also be devolved to such centres from national capitals.

In contrast, developments in the insurance industry may lead to a bypassing of regional centres. Brokers previously reliant upon the back-up of regional offices for the preparation of quotations are now able to 'key in' details of particular proposals into central computers, thus reducing the need for a network of regional offices.

In other producer services similar forces for both centralization and decentralization are also at work. The exact nature of the impact of new information technology will depend on the corporate organization of the sectors concerned, particularly the balance between large, multi-region and small, local companies. Many business service activities have traditionally been the preserve of small and medium-sized enterprises serving local markets. However, the 1960s and 1970s have witnessed an increasing concentration of ownership within key business services such as accountancy and management consultancy. For example, a survey of ten business service sectors (accountants, finance companies, insurance brokers, solicitors, advertising agencies, computer bureaux, architects, consultant engineers and management consultants) in three provincial regions of the United Kingdom has revealed that 67 per cent of establishments are branch offices and 71 per cent had their headquarters in London (Goddard and Marshall, 1983). Such concentration has occurred through the internal expansion of companies and the establishment of branch offices in various locations, and external expansion through the acquisition of previously independent local firms.

Developments in telecommunications have no doubt facilitated this process of ownership concentration and the geographical expansion of services by large companies. While this branch expansion has produced additional employment and improved the quality of services available in some cities, it may also have been at the expense of indigenous service industries. In so far as a more widespread use of new information technology results in greater dispersal of business service firms from capital cities, the *net* effect on employment in other cities may not be positive. In contrast to mobile manufacturing firms, service enterprises are likely to compete directly in local markets, a fact that is often overlooked in policies designed to stimulate mobility in the service sector.

5. Conclusions for urban and regional policy

The discussion presented in this chapter indicates how difficult it is to isolate the independent effects of technological change on economic development. This problem is compounded when one moves from the national and international levels to the regional and urban levels. At these scales, the inertia of the built environment, coupled with the constraints imposed by existing organizational, industrial and institutional structures become a major consideration. Scenarios which seem reasonable in the abstract context of the national economy, or even in the context of the individual industry or company, can fall apart when projected on the specific realities of particular urban or regional economies (Goddard and Thwaites, 1980).

In consequence, the speed of diffusion of new technologies is often grossly overestimated. In addition, the secondary effects of technological change are often overlooked—for example, the possibility that by increasing the opportunities for interaction within society the use of telecommunications can generate more travel than it substitutes for. Equally, the distributional effects of technological change are overlooked by neglect of the spatial diffusion process. Who gains and who loses is intimately related to the pattern of diffusion of technology through space and time, related, of course, to the way in which existing institutions adjust. Many scholars have argued that the economic crisis facing Europe today arises because the institutional and physical fabric of society, with its high degree of inertia, is unable to change rapidly in order to exploit the outstanding cost and productivity advantages of the new technology (Freeman, 1984; Perez, 1983). Necessary institutional changes embrace education and training, industrial relations, management and corporate structure, capital markets and public investments. While many of the necessary changes require action at the national level, such initiatives may 'run into the sand' unless complementary policies are pursued at the urban and regional scale. In this context there is a considerable role for information technology strategies which mobilize resources and deliver policy at the regional level; within such regional strategies the regional capital city must clearly play a central articulating role.

A regional information technology strategy would need to be overseen by a powerful body containing representatives of all the potentially interested parties (e.g. the PTT, industry and commerce, trade unions and educationalists). A well-formulated strategy is needed to monitor developments, bid for resources, encourage and co-ordinate policy initiatives. The following activities are indicative of the tasks involved:

(1) Keep abreast of infrastructure investments and new developments in technology and services. Lobby for investment and innovative projects (such as fibre optic experiments and cable-TV networks), so helping to ensure the area is well served and develops expertise in information technology (e.g. interactive cable services).

(2) Support the marketing of telecommunications services: publicity and awareness exercises, demonstration projects and resource centres.

(3) Lobby for tariff reductions to facilitate access to core regions.

(4) Co-ordinate and give greater publicity to existing initiatives, such as technology transfer; seek out and support new initiatives and projects, such as regional databases on goods and services, tourist facilities and export opportunities. Bid for new projects and 'top-up' resources to support innovation, in addition to those provided through national or European policies which are not regionally differentiated.

(5) Set up centres/agencies able to provide, at an accessible 'shop-front' location, a wide range of assistance on the adoption of information technology. This could include access to technical databases, access to facsimile, telex, etc, on a bureau basis also offering use and 'hands on' experience of automated office systems, microcomputers, etc. Such an activity would need fieldworker staff, especially to deal with technology transfer activities within manufacturing industry.

(6) Demonstrations of telecommunications networking and the inclusion of its potential within promotional efforts to encourage inward investment. It is necessary to show enterprises the capabilities of telecommunications as an aid to branch managements and as a way of facilitating decentralization of some functions.

(7) So far as information technology production is concerned, there is a need to identify specialist sectors which build on the industrial production strengths of the area and focus efforts on establishing these via indigenous development and inward investment. However, as far as possible competitive bidding for inward investment should be avoided.

(8) Co-ordinate information on current and future skill requirements; support/lobby for training and retraining programmes.

Obviously, not all these proposals will be relevant to each region, and some regions may have specific needs which might be unsuitable for inclusion within an information technology strategy. However, it is worth noting that one of the strengths of an information technology strategy is that the technology is of such general applicability, and has such widespread impacts, that it can draw in a multiplicity of issues and concerns, cutting across boundaries of conventional policy areas.

The role of the provincial cities in carrying out these strategies in the less favoured and peripheral regions is undoubtedly crucial. They are the focal points of their regions, best able to take a leading role in the diffusion of technological awareness and change. Above all, they have the institutional infrastructures capable of supporting policies to stimulate information technology. Although these institutional resources have often become depleted—notably through the erosion of financial and business services—the provincial cities of Europe generally have higher education institutions, chambers of commerce, regional headquarters of the tele-communications services, local government and branches of central government. Many have regional development and promotion organizations as well. To implement an IT strategy successfully, these institutions would have to work together and support each other.

These cities have some particular characteristics especially relevant to information technology. Certainly their educational institutions have much to offer information technology producers and users such as quali-fied manpower, training and R & D support. In many cases there remains considerable scope for greater involvement of these institutions in their local economies. The cities can offer environments conducive to the relocation of clerical work which has become more mobile with the spread of distance-independent data transmission services. Provincial cities often have plentiful supplies of clerical workers, low-cost accom-modation and other services. The cities are very much more likely to be 'cabled' than small towns (let alone rural areas) and hence may have the opportunity of participating in the provision of Value Added Net-work Services, including city-wide information and interactive services between enterprises.

In a variety of situations the provincial cities would have to play a key role in establishing, leading and implementing the kind of strategy outlined above. They must show, by example, what information tech-nology can offer and so set in motion a process of diffusion into their surrounding regions. The cities must service the activities of their hinter-lands—such as agriculture or tourism—acting as brokers between hinter-lands and markets elsewhere. In some ways new information technology may weaken the position of the cities, as hardware production favours green-field locations and 'remote working' strengthens the attraction of locations outside the city. This makes it all the more imperative that strategies must be regional and the mutual dependence of city and region recognized.

Information technology is not a universal panacea for all the problems of lagging regions or old industrial areas. But it can offer them some new opportunities—in contrast to the apparently unavoidable and negative impacts on existing employment levels. A coherently structured strategy, sensitive to a region's specific needs, administered within the region— and drawing heavily on the resources of the provincial cities—could enable depressed regions to maximize the potential benefits of this new technology. The so-called 'information technology revolution' requires a positive response; it cannot be ignored.

References

Clark, D. (1978). *The Spatial Impact of Telecommunications*, DoE Research Report 24 (London).

EEC (1982). *Information Technology and Job Creation in Europe, FAST 1* (Brussels).

Freeman, C. (1984). 'Keynes or Kondratiev: how can we get back to full employment', in P. Marstrand (ed.), *Technology and the future of work* (London, Frances Pinter).

Freeman, C., Clarke, J. and Soete, L. (1982). *Unemployment and Technical Innovation: A Study of Long Waves* (London, Frances Pinter).

Gershuny, J. I. and Miles, I. D. (1983). *The New Service Economy* (London, Frances Pinter).

Gibbs, D. G. and Thwaites, A. T. (1983). 'The Stimulation of Economic Activity in Local Enterprise through the provision of Technical and Commercial Information', Final Report, October 1983, Centre for Urban and Regional Development Studies, University of Newcastle upon Tyne.

Goddard, J. B. (1983). 'Industrial Innovation and Regional Economic Development in Britain', in Hamilton and Linge (eds), *Spatial Analysis, Industry and the Industrial Environment* (London, Wiley).

Goddard, J. B. and Marshall, J. N. (1983). 'The future for offices in the City Centre', in R. L. Davies and A. G. Champion (eds), *The future of the City Centre* (London, Academic Press).

Goddard, J. B. and Pye, R. (1977). 'Telecommunications and Office Location', *Regional Studies*, 11, 19–30.

Goddard, J. B. and Thwaites, A. T. (1980). *Technological Change in the Inner City*, Working Paper No. 4, Inner Cities in Context Working Party, Social Science Research Council, London.

Gottman, J. (1966). 'Why the Skyscraper? ', *Geographical Review*, 15, 190–212.

Langdale, J. (1983). 'Competition in the United States Long-distance Telecommunications Industry', *Regional Studies* 17, 393–410.

Marstrand, P. (ed.) (1984). *New Technology and the Future of Work* (London, Frances Pinter).

Meier, R. L. (1962). *A Communications Theory of Urban Growth* (Cambridge, Mass., MIT Press).

OECD (1981). *Information Activities and Electronic Telecommunications Technology* (Paris).

Peck, F. and Townsend, A. (1984). 'Contrasting Experience of Recession and Spatial Restructuring: British Shipbuilders, Plessey and Metal Box', *Regional Studies* 18, 319–38.

Perez, C. (1983). 'Structural change and the assimilation of new technologies in the economic and social systems', *Futures*.

Pool, I. de Solla (1978). *The Social History of the Telephone* (Cambridge, Mass., MIT Press).

Prais, S. (1976). *Industrial Concentration* (Cambridge, Cambridge University Press).

Schumpeter, J. A. (1939). *Business Cycles* (New York, McGraw-Hill).

Webber, M. W. (1964). 'The urban place and non-place urban realm', in D. L. Foley (ed.), *Explorations into Urban Structure* (Philadelphia, University of Pennsylvania Press).

Index